A CONCISE HISTORY OF MINING

Pergamon Titles of Related Interest

Keller ELECTRICAL METHODS OF GEOPHYSICAL PROSPECTING
Wills MINERALS PROCESSING TECHNOLOGY
Sideri MINING FOR DEVELOPMENT IN THE THIRD WORLD
Carman OBSTACLES TO MINERAL DEVELOPMENT

Related Journals*

JOURNAL OF TERRAMECHANICS
RECLAMATION REVIEW

*Free specimen copies available upon request.

A CONCISE HISTORY OF MINING

Cedric E. Gregory

Pergamon Press

New York • Oxford • Toronto • Sydney • Paris • Frankfurt

Pergamon Press Offices:

U.S.A.	Pergamon Press Inc., Maxwell House, Fairview Park, Elmsford, New York 10523, U.S.A.
U.K.	Pergamon Press Ltd., Headington Hill Hall, Oxford OX3 0BW, England
CANADA	Pergamon of Canada, Ltd., Suite 104, 150 Consumers Road, Willowdale, Ontario M2J 1P9, Canada
AUSTRALIA	Pergamon Press (Aust.) Pty. Ltd., P.O. Box 544, Potts Point, NSW 2011, Australia
FRANCE	Pergamon Press SARL, 24 rue des Ecoles, 75240 Paris, Cedex 05, France
FEDERAL REPUBLIC OF GERMANY	Pergamon Press GmbH, Hammerweg 6, Postfach 1305, 6242 Kronberg/Taunus, Federal Republic of Germany

Second printing, 1982

Library of Congress Cataloging in Publication Data

Gregory, Cedric Erroll.
A concise history of mining.

Bibliography: p.
Includes index.
1. Mining engineering—History. I. Title.
TN15.G73 1980 622'.09 80-13925
ISBN 0-08-023882-3

Printed in the United States of America

To the cream of the world's society,
the salt of the earth:
those who live in mining camps

Ancient surface workings of the Ophir Mine worked by King Solomon (c. 900 B.C.) at Mahd adh Dhahag, Saudi Arabia. The author visited this famous old gold mining area in 1979.

Contents

List of Tables

List of Figures and Plates

Foreword

It is particularly fortunate that Professor Gregory has chosen to write the history of mining at this time. The history of mining has been the history of civilization and the early development of nations. Where minerals and mining existed, they provided ingredients for weapons, wealth, and world power.

Today as newly created nations are developing, the importance of minerals and mining again has emerged. The geopolitics of mineral and fuel deposits has highlighted the new wealth of Saudi Arabia and the needs of South Korea. Since history is a source of knowledge and perspective, Prof. Gregory's text is particularly timely to accompany this period of developing countries and the North-South dialogue.

It is fortunate that one of Prof. Gregory's stature and experience is moved to write on this subject as he prepares to retire.

Walter R. Hibbard
University Distinguished Professor
University of Idaho

Preface and Acknowledgments

My purpose in writing this little book is to provide students with a suitable text for a service course offered by the Department of Mining Engineering in the University of Idaho as a humanities elective for students majoring in a wide range of study programs across the campus. This course, originated by my colleague, Dr. Christopher Haycocks, has proved to be extremely attractive, mainly because it has inspired students to develop a genuine interest in the subject.

Nevertheless, since John Temple's Mining - An International History has ceased to be available, we have felt that students are at a disadvantage. By and large, books on general history are notoriously difficult to read and understand, except only for those who already know their history. History books seem to deal predominantly with political aspects, relieved at intervals by a few soul-tearing episodes, such as the "charring of the cookies" by King Alfred.

For this reason, history has unjustly earned a reputation of being the "driest" and among the least interesting of subjects of study. It need not be so. There are many important aspects of history other than the political or the religious.

As for the history of mineral production, which has had a tremendous impact on the development of the human race over a period of 50,000 years, it is difficult not to agree with Rickard, who states:

> Mining has played a bigger part in the development of civilization than is usually conceded by the historian or even surmised by the ordinary citizen. The reason for the failure to appreciate the importance of mining . . . is the historian's lack of familiarity with the technique of mining and metallurgy, and his [apparent] aversion from discussing such matters.

Yet, there is no dearth of books published on the subject of mining history; but although some of these are excellent treatises, most are too voluminous, or rather ponderous, or unduly dramatic or nostalgic, or too specialized, in that they cover only certain time ranges, limited geographic regions, or the mining of particular minerals.

On the other hand, we have felt a need for an easy-to-read text that in simple language embraces the broad canvas of mining history from the Old Stone Age to the present, and a wide range of mineral production across the globe, though limited necessarily to a reasonably selective coverage. These aims can be achieved, for a course to be delivered in a single quarter or semester, only if a condensed skate-across-the-board approach is adopted. There is no possibility of a deeper study in the time available.

History, of course, cannot be studied objectively except within a framework of time, labelled by dates. It should be understood, however, that the onset of some ages did not occur evenly over the known world. For instance, the Bronze Age commenced a thousand years in western Asia before it reached Europe, and in fact some centuries before its effects were seen in Egypt. Apart from this, there is some disparity among the different authorities in quoting dates of events in more recent years. This is perhaps because of the lead time involved in establishing a producing mine, usually many years after the threshold discovery year, especially in those cases where the development program becomes shelved for some years while awaiting the determination of a suitable metallurgical method of treating the ore.

In order to present the subject matter in a readable, understandable manner, I have purposely departed from accepted practice by eschewing the use of interpolated references and footnotes. For those who would delve deeper, an extensive bibliography appears at the back of the book.

Since this is a historical study of a specialized sector of civilization, it is necessary for the reader first to become familiar with the technology and terminology of the particular field. Part I, therefore, introduces in simplistic terms a general discourse on mining techniques. Later sections deal with the historical aspects of mining.

In preparing this text, I have drawn heavily on my own personal experience as a hard-rock miner (three years), as a mining engineer, as a mine manager, and as a mining professor. My travels in 93 countries have included visits underground in mines on all continents: from Swedish Lapland through Europe to South Africa, Australia, Southeast Asia, and North America. Some of these mines are of rare vintage, such as the Rammelsberg (established in 968 A.D.), and the Rio Tinto Mine, worked originally by the Phoenicians.

Although this book has been prepared primarily for the university student, its simplistic, uncluttered style should also attract the interest of the general reader; and after all, there are cogent reasons why the community should become much more interested in mining activities, without which our standard of living could not possibly endure.

The book is written in American spelling. Readers should also observe that many of the geographical figures (maps) do not purport to reflect current geopolitical state boundaries or nomenclature. Many of these have changed quite radically and frequently over the years of recorded history, and are still being changed.

Armed with this textbook, any professor of mining engineering, or even of the geological sciences, in any university should find no difficulty in introducing and teaching a service course in the history of mining as a humanities subject to students majoring in any discipline. Naturally, the standard of the presentation will depend upon the nature and extent of the professor's personal field experience.

I am especially indebted to my colleague, Dr. Christopher Haycocks, for reviewing the manuscript and for his valued suggestions; to Dr. J. Richard Lucas, for his warm support and encouragement; to Dr. Walter R. Hibbard, Jr., University Distinguished Professor, for his sincere interest and staunch support, and for writing the Foreword; to Mr. L. P. Johnson III, for his ever ready assistance; to Mr. Wayne Slusser, for his assistance with photographs; to Mrs. Susan M. Breeds, for painstaking checking of the manuscript; and to the staff of the Learning Resources Center for preparation of the diagrams.

Grateful appreciation is also accorded the following publishers who have kindly granted permission for quotations and excerpts:

Penguin Books Ltd. (Metal Mining by J.B. Richardson);

McGraw-Hill Book Company (Mining in the New World by Carlos Prieto);

my sincere friend, Dr. Wolfgang Paul (Mining Lore);

Ernest Benn Limited (Mining - An International History by Temple), also for figures 11.1, 11.2, 12.1, and 12.3;

Arizona Historical Society (Early Underground Mine Lamps by Pohs), also for plates 12.5, 19.1, 19.2, and 19.3;

Michigan Technological University (*Stones of Destiny* by Poss);

Hicks Smith & Sons Pty. Ltd. (*An Introduction to Mining* by Thomas), also for figures 4.8, 4.9, 4.10 4.11, 4.12; and

Westfalia Luenen, also for plates 23.1 and 23.2.

However, I personally prepared the whole of the manuscript; therefore, for any inconsistencies and omissions, the buck must stay with me.

I

What is Mining?

1 Introduction

The study of mining history is not very satisfying unless we have some background knowledge of mining and its associated disciplines. So first of all, we are confronted with questions like these:

What is "mining"?

What is the nature of the operations involved in the various fields of mining activity?

Well, of course, we could start with a definition:

> Mining is the process of extracting minerals of economic value from the earth's crust for the benefit of mankind.

Here we can assume that the earth's crust refers to the outer surface of the earth extending to a depth of about 20 miles, including the oceans, lakes and rivers, and the surrounding atmosphere.

But before we proceed further, let us see where mining fits into our economic system. What part does mining play in contributing to our national output of goods and services (the gross national product)?

THE STRUCTURE OF THE UNITED STATES ECONOMY

As with most free-enterprise nations, the structural framework of our industrial economy embraces three different groups:

1. <u>primary industry</u>, which includes agriculture (covering also forestry and fisheries), and mining. These are the only two basic primary industries, in that the goods they

produce are derived from the earth's crust, including the oceans. In turn, the products of agriculture and mining become the raw materials for secondary industry. (Some agricultural products, such as fresh vegetables, are used directly by consumers.)

2. secondary industry, in which the raw materials used as feedstock are the products of the two primary industries (either of domestic or foreign origin), and processed by various manufacturing techniques. As an example, agricultural products are typically fed to food-processing plants, and to the paper-making, the building and construction industries, and other market outlets. Similarly, the mineral products of the mining industry are utilized in a number of vital ways, by a wide range of processing industries. These cover power generation, steel production and fabrication, nonferrous metal fabrication, fertilizer production, and various processing techniques for the chemical, pharmaceutical, building, decorating, communication, cosmetic, defense, and electronics industries. These techniques embrace the manufacture of all types of machine assemblies, such as farm implements, automobiles, appliances, airplanes, ships, and many other items of machinery and tools.

These lists are by no means exhaustive. They are mentioned merely to give some sort of indication as to where secondary (manufacturing) industry derives its raw material feedstocks: from the agricultural and mining industries.

But, of course, no complex economy could operate with materials produced and stored at the factory door. In order to distribute these manufactured products among the consuming sectors of the public, we have a third component of the economy. It produces services.

3. tertiary industry involves: a) the sales and distribution (wholesale and retail) sectors of industry; b) the various transportation sectors, by land, sea, and air; c) professional services, including health, law, education, and religion; d) services of the various levels of government; e) maintenance services, such as the various repair trades, laundries, gas stations, hairdressing, and the like; f) the entertainment industry; and among others, g) financial, banking, and insurance services.

In our complex industrial economy, it is obvious that none of the three industry groups (primary, secondary, and tertiary) could be effective if operating alone. All three groups are interdependent in a highly developed society.

It is true that some early tribal communities were able to exist, as "peasant" economies, on agricultural output alone. But today, in most industrialized nations, even agriculture depends upon mining for fertilizers, chemical sprays, and machinery. To be sure, mining also depends upon agriculture for food, for many housing materials, and for clothing made from natural fibers (three of the prime necessities of life), as do the members of all industry groups, and humanity in general.

Nevertheless, although humanity may be able to exist (barely) on food, shelter, and some clothing in a peasant economy, provided air and water are also available, the existence and maintenance of any industrialized society involves every member of that society in a high level of dependence upon mineral production (i.e., on mining activities). Therefore, it is very important for all members of our society to become actively interested in their mining industry.

It is interesting to note that at least one food item (salt) is recovered by mining techniques; so also are the raw materials for many medicines, for most plastics, and for all surgical instruments.

Now we have two more questions: What is a mineral and what is a mineral of economic value?

We will deal with these in a simplistic way because it must be assumed that readers have no knowledge of general geology and very little of basic chemistry.

So now we must come to the point of distinguishing between chemical elements, minerals, and rocks.

Chemical Elements

Chemical elements are particles of matter existing as distinctive atoms. Each specific element has its own atomic weight and other characteristics such as specific gravity (density). All these elements are related to one another in a framework of characteristics known as Mendeléev's Periodic Table of the Elements. There are about 104 such elements that have been isolated and recognized in the laboratory. But most of them are found in the field in the combined form: combined with other elements.

Some elements occur in the natural state in greater quantities than others. For instance, 99 percent of the earth's crust consists of only 8 such elements, as set out in the following, with the chemical symbol of the element shown in parentheses:

Oxygen	(O)	47.0%
Silicon	(Si)	28.0%
Aluminum	(Al)	8.0%
Iron	(Fe)	5.0%
Calcium	(Ca)	3.6%
Sodium	(Na)	2.8%
Potassium	(K)	2.6%
Magnesium	(Mg)	2.1%

This means that all other 96 elements are relatively scarce in that they represent only 1 percent by weight of the earth's crust. In fact, some of the more important metals (apart from aluminum and iron) that we must rely upon so greatly to sustain our standard of living represent very minute proportions of the earth's crust, e.g., copper (Cu), 0.0045 percent; lead (Pb) 0.00015 percent; gold (Au), 0.0000007 percent.

It will be seen that these and other important metals are dispersed very thinly in the earth's crust. At such low concentrations, we cannot afford to recover them (mine them) for the benefit of society, any more than we can afford to mine aluminum and iron at average concentrations of 8 and 5 percent, respectively.

Well, how do we get over this problem? How is it that these metals are being mined? The answer to these questions is about to be unfolded.

But first let us look at another aspect.

With few exceptions, these metals do not occur in the earth's crust as elements. More usually they are found combined or compounded with other elements as we shall see; not the sort of chemical compounds that we find prepared artificially in the laboratory, but chemical compounds that are found in the natural state in the earth's crust. These are called minerals, which we will deal with next.

But first we will refer to those few exceptional cases of elements that occur "native," i.e., in the natural state. There are very few. Some are:

Copper	(Cu)
Silver	(Ag)
Gold	(Au)
Platinum	(Pt)
Sulphur	(S)
Graphite	(C)
Diamond	(C)

and the elemental gases of the atmosphere (O, N, etc.).

Minerals

We may define minerals as "naturally occurring inorganic chemical compounds with relatively precise chemical compositions and distinctive physical properties." In a simple way, we may think of them as aggregates of chemical elements, although this may not be a scientifically acceptable term. There are about 2,000 catalogued species.

Each particular mineral has its own mineralogical name, which is quite distinct from the name of the artificially prepared chemical compound that you find in the laboratory.

Although the chemical composition may be the same, the physical characteristics are quite different. For instance, lead sulphide, a chemical compound of lead and sulphur, with the chemical symbol of PbS, is generally available in the laboratory as an amorphous powder. It has quite different physical properties from galena (PbS), the name of the naturally occurring mineral, often in crystallized or crystalline form. One particular mineral deserves special mention perhaps because it is of widespread occurrence. It is water, a chemical compound of hydrogen and oxygen.

There are two exceptions to our general definition of a mineral as expressed above: 1) there are some minerals that occur native, or uncombined, such as native copper, native gold, etc. These do not quite fit our definition; and 2) there are other naturally occurring materials that are not inorganic in origin and have no precise chemical composition and no definite physical characteristics. But for convenience, we consider them as minerals. They are naturally occurring "organic" hydrocarbons, such as coal and petroleum.

Minerals fall into two categories for the purpose of our study: common rock-forming minerals and "useful minerals" (minerals of economic value). When sufficiently localized or concentrated by nature in what are known as mineral deposits, these latter are said to be high enough in grade to warrant extraction by mining.

Rocks

Rocks are classified as igneous, sedimentary, or metamorphic; but these distinctions need not concern us here. Whereas minerals may be considered as aggregates of chemical elements, so rocks are really aggregates of minerals. They contain minerals in varying proportions, and therefore they have no definite chemical composition.

In general, rocks are of no particular economic importance, except in the case of those with desirable physical characteristics, such as construction stone, building stone, and monumental stone. Such rocks are mined for these purposes.

Two mono-mineralic types of rock should be mentioned. One is generally referred to as limestone. It is basically a mineral (calcite, $CaCO_3$) in a crude impure form, but it occurs in such widespread deposits that it is generally regarded as a sedimentary rock. Limestone is mined as a very important raw material for cement manufacture, and for other purposes. Another variety of calcite, occurring in large deposits, is marble, generally regarded as a rock, but usually pure enough to be rated as a mineral. The other main mono-mineralic rock is known as sandstone; it is really a large mass of consolidated crude quartz grains.

Coal is perhaps more accurately regarded as a carbonaceous rock as is oil shale. But in no way can we consider liquid petroleum (crude oil) as a rock, so we now accept these three main organic hydrocarbon materials as special types of minerals, as described earlier.

We now have a reasonably clear idea of the composition of the lithosphere (the solid part of the earth) in terms of its separate classes of constituents. From now on we will omit serious consideration of the gases in the atmosphere, or the waters of the oceans, lakes, and rivers, although these continue to play their part in the slow, natural geological processes of weathering of rocks, which they convert into soil and other sediments. But they are not of major direct interest as economic materials to be mined.

The real economic targets of mining activity are about 100 of the 2,000 or more catalogued mineral species, including the native minerals, the hydrocarbon minerals, and the few types of economic rocks already mentioned.

But in terms of our previous discussion, how can we consider mining these materials if they are so widely dispersed among the barren, uneconomic rocks? This is the next important point to consider.

Actually, the dispersion of minerals throughout the earth's crust is not an even, uniform phenomenon. In places we find useful minerals concentrated or localized by nature in various modes, as we shall see later. For instance, fractures in the rock mass may serve as channels for the deposition of mineral-bearing solutions, from which the minerals are then precipitated to form what we call veins. If the fractures are wide zones of weakness in the rock, we often find wider deposits called lodes. There are other modes in which mineral deposits occur.

Under our private enterprise system, only those mineral deposits will be mined that are sufficiently concentrated by nature (i.e., of high enough grade) to be extracted at a profit. And mineral deposits can be mined only where such deposits are found to exist, however inconvenient the location.

2 Mineral Deposits

CLASSIFICATION OF ECONOMIC MINERALS

There is a broad range of "useful" minerals that, after mining and processing, are customarily used as raw materials for the secondary or tertiary industries. This range of economic minerals can be classified under the following headings. The list is by no means exhaustive.

1. Minerals that yield metals
   Precious metals: gold, silver, platinum.
   Base metals: copper, lead, zinc, tin.
   Steel industry metals: iron, nickel, chromium, manganese, molybdenum, tungsten, vanadium.
   Light metals: aluminum, magnesium.
   Electronic industry metals: cadmium, bismuth, germanium.
   Radioactive metals: uranium, radium.
2. Nonmetallic minerals
   Insulating materials: mica, asbestos.
   Refractory materials: silica, alumina, zircon, graphite.
   Abrasives and gems: corundum, emery, garnet, diamond, topaz, emerald, sapphire.
   General industrial minerals: phosphate rock, rock salt, limestone, barite, borates, felspars, gypsum, potash, trona, clays, magnesite, sulphur.
3. Fuel minerals
   Solid fuels: anthracite, coal, lignite, oil shale.
   Fluid fuels: petroleum oil, natural gas.

CLASSES OF MINERAL OCCURRENCE

We may rate the relative possibility of economically mining deposits of useful minerals in three broad classes as follows:

1. Natural concentrations of economically useful minerals. They may be of relatively high or low grade, but nevertheless capable of being mined profitably. Some of these we know about, but there are obviously many other such deposits that are still waiting to be discovered. This is the class that represents our present mining target.

2. Mineral concentrations of submarginal grade, whether known or as yet undiscovered. At present costs and levels of technology and mineral market prices, this class is generally too low in grade to justify mining operations. In future years, as class (1) occurrences become exhausted, this class may fit into our target.

3. Dispersed minerals throughout the rocks of the earth's crust. These may be (a) common rocks forming minerals, such as quartz, felspars, and silicate minerals, or (b) potentially useful minerals. But these latter are too widely dispersed and low in grade, and therefore of no economic value at prices ruling at present or in the foreseeable future. Therefore, it will be seen that the production of minerals of economic value by mining techniques must be confined to those covered by class (1) considerations.

Our next approach is to study the various modes in which these localized mineral deposits occur. But before we go too far, we should acquaint ourselves with the particular terminology used in the mining industry. Definitions of mining terms are listed in the Glossary at the end of the book.

MODES OF OCCURRENCE OF MINERAL DEPOSITS

There are several accepted methods of classification of economic mineral deposits that have been localized or concentrated by nature. But we will classify them according to the mode or form of the deposit. Three modes of occurrence are recognized.

Veins are simply fissures, fractures, or fault planes in a rock mass that have been filled by precipitation of minerals from solutions or by injection of magmatic mineral material from the depths of the earth. Lodes, which are in zones of fractured rock, are similarly filled. Because these fissures and fractures generally occur radially, most veins and lodes have a near-vertical habit. They may be of several hundred feet in strike length, and the width is generally variable. Veins may bulge or pinch out over quite short distances in strike or depth. Lodes are usually wider and of lower grade material.

Bedded deposits are typified by coal seams which are interbedded between the layers of stratified sedimentary rocks. Coal seams were formed from plant growth followed by a gradual sinking of the land below sea or lake level. The plant life then decayed and peat was formed. Sandy and muddy sediments settling through the water over periods of thousands of years became highly compressed and gradually formed sedimentary rocks. The pressure from these overlying sedimentary rocks converted the peat into lignite and much later into coal. Seams of coal can be seen outcropping between beds of sedimentary rock in many of the highway cuts in the eastern United States. Most of our coal was formed from plants that grew about 250 million years ago. Most coal seams are horizontal beds.

Replacement deposits also occur mainly in a more or less horizontal habit. They are metalliferous deposits that were formed by the dissolving away of sedimentary beds of limestone rock by mineral-bearing solutions which partly replaced the rock.

In between coal and metalliferous mineral deposits we have a wide range of industrial mineral (generally nonmetallic) deposits. Some of these deposits occur as veins, and others as stratified (bedded) deposits.

Bedded deposits may extend over large areas. Their lateral extent is usually much more widespread and continuous than metalliferous nonferrous deposits. They contain minerals such as coal, iron ore, salt, potash, gypsum, and phosphate deposits. Limestone, required for cement manufacture and for use as a smelter flux, is also mined from bedded deposits.

When the relatively soft industrial minerals are mined underground in bedded deposits, coal mine practice is generally followed. However, metalliferous (replacement) deposits in the horizontal mode are in hard rock and need to be mined with explosives.

GRADES OF ORE

As will be seen from the definitions listed in the Glossary, the grade of an ore deposit is the equivalent average weight of metal contained in a ton of the ore, even though the metal is in one or more of its mineral forms. The ore grade is, therefore, of considerable economic significance because the metal recovered from every ton of ore must yield enough revenue to cover the cost of mining, milling, and smelting that ton of ore (plus overheads); otherwise no profit will ensue.

This means that a great deal of attention must be paid to ensure that only ore of greater than the minimum economic payable grade is mined. This value is known as the "cut-off" grade: a grade that is marginal between a profit and a loss. Submarginal-grade ores can be mined, but only at a loss. The lower the unit working costs, the lower the cut-off grade, and therefore the greater amount of previously submarginal-grade ore that can now be profitably mined. This represents an increase in national wealth. On the other hand, government taxes add to costs, raise cut-off grades, and therefore reduce national wealth.

Because various methods of mining necessarily entail different unit costs, it follows that the lowest cut-off grade will be associated with the least expensive mining method, e.g., open cut work (surface mining). If, however, the particular mineral deposit is situated well below the surface, it will need to be worked by one or other of the applicable underground mining methods, which are more expensive. In this case, the minimum profitable cut-off grade will necessarily be higher. Nevertheless, because the various metals are priced at different dollar values on the world markets, and for other reasons, it also follows that ores containing single metals will have widely different cut-off values.

Typical minimum profitable cut-off grades for some of the metals are as follows. They are expressed as for open cut mining.

Antimony	4.5%
Copper	0.5%
Chromium	35.0%
Gold	0.15 oz/ton
Iron	28.0%
Lead	2.0%
Manganese	28.0%
Nickel	0.5%
Platinum	0.12 oz/ton
Silver	1.0 oz/ton
Tin	0.45%
Zinc	3.0%

The foregoing discussion applies to what are known as "straight" ores: ores containing minerals of a single metal. It is interesting to note that the cut-off grade of straight gold ores is about one-sixth of a troy ounce per ton. This is about the size of a five-cent piece, recovered from 12 cubic feet of ore (about one ton).

However, many metalliferous minerals occur in close association with one another, and therefore several marketable products may be recovered at little extra cost from the same ton of ore. For instance, we find that lead and zinc sulphide minerals frequently occur in the same ore body, and some silver is also present. This means that a ton of this ore will yield greater revenue and the mean cut-off grade can, therefore, be reduced. Other complex ores may also contain such metals (in mineral form) as copper, antimony, molybdenum, and pyrite.

There are also other interesting features. It so happens that most mineral deposits containing high-grade ore are small in dimensions. These must be worked on a small scale by relatively costly manual methods.

On the other hand, many massive deposits (such as wide lodes) of low-grade ores have been found in recent years. These can now be worked profitably by mass-production methods, mostly on the surface but also underground, because more space is available for mechanized equipment.

3 Three Aspects of Mining Activity

Figure 3.1 shows the main distinctions between the mining and processing of metalliferous minerals, industrial minerals, and fuel minerals (coal), up to the point where they reach the market. We see that metalliferous minerals occur in ore deposits - in veins, reefs, or lodes - in a near-vertical habit, or as bedded deposits in horizontal formations. In each case, the ore and the enclosing rock are very hard and can be mined only with the use of explosives.

In a typical ore body, the valuable minerals occur in a groundmass or matrix of other worthless minerals called gangue. In other words, the gangue may represent 95 percent or more of the total amount of ore mined. The valuable minerals are often disseminated in tiny blebs or veinlets throughout the gangue.

The main problem is to beneficiate the ore by separating and discarding the gangue as early as possible in a mill (concentrator, or treatment plant) set up on the surface. This can be done after first crushing and grinding into very fine particles the lumps of ore delivered from the mine. The actual separation is then done by one of the many processing methods, generally referred to as concentration because the separation and discarding of the gangue minerals as tailings leaves behind a 'concentrate': a much smaller bulk of material containing practically all the valuable mineral. This separation process is not fully achieved in practice because a small fraction of the valuable mineral is usually lost in the tailing. But the concentrate is still in the form of a mineral which has no sale value in the general market. The next step is to extract the metal from the mineral concentrate.

The processing unit in which metal extraction is performed is typically called a smelter established on the mine site, or perhaps at some industrial center or port, where fuel

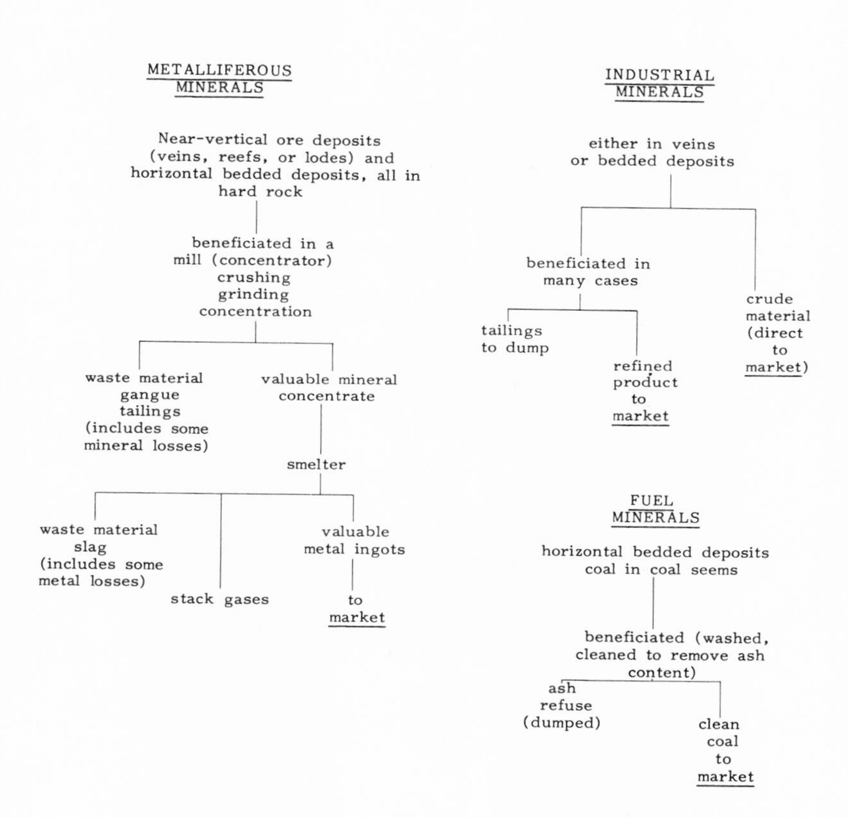

Fig. 3.1. Three aspects of mining activity.

and electric power are available at a reasonable cost. In the smelter, the concentrate, along with fluxing agents and coke fuel, are fed to a blast furnace. In time, the mineral concentrate is reduced to a molten metal collecting at the bottom of the hearth. Some of the impurities collect and float above this as a fluid slag, and others pass along the flues and up the stack as gases (typically sulphur dioxide) and fine dust, which are now being recovered or purified before being released to the atmosphere. At intervals, the molten slag is tapped and run off into a suitable receptacle, granulated into small pieces with a stream of water, and stored or dumped for sale as rubble or road metal. Similarly, the molten metal bath is tapped at intervals and run into the specially shaped ingot molds of a casting machine. On cooling, the solid ingots of metal are shipped to market. Again, not all the metal contents of the concentrate are recovered.

This is a necessarily simplistic description of a typical concentration and smelting process for the recovery of metal ingots from a metalliferous ore. Many variations occur in practice, and the procedures are specific for each class of metal.

There is a wide range of industrial minerals mined for a myriad of industrial end uses. They may occur in near-vertical vein deposits, or in bedded deposits, or in other modes. The run-of-mine material delivered to the surface may be marketed directly where the end use calls for a crude cheap product (as with barite, used as a drilling mud in oil well exploration work); or it is beneficiated to remove the impurities where a high technical grade product is required. The beneficiation plant and methods are based upon the particular chemical process reactions required to remove the impurities before the cleaned product is delivered to the market. The impurities are seldom more than 20 percent of the mined material.

With fuel minerals the proportion of worthless material, generally called ash, is usually no more than 25 percent by weight. This represents shale and earthy materials (sometimes pyrite) which are unavoidably mined along with the coal.

The valuable constituents of coal are the combustible carbon and volatile gases that together represent the heating potential, or calorific value, of the particular coal. The incombustible ash obviously makes no such contribution. If the coal needs to be transported great distances to market, then it will certainly be advantageous first to remove the ash in order to economize on freight charges.

Ash is therefore removed by a beneficiation process, in which the run-of-mine coal is beneficiated (cleaned, or washed). This is performed in a cleaning plant (or coal washery) erected on the surface. The ash is dumped as refuse or slag, and the cleaned coal is loaded into railroad cars or river barges for dispatch to the market.

Because coal has no definite chemical composition like regular minerals of inorganic origin, it cannot readily be evaluated in terms of "grade." It is usual to evaluate (or to rank) a coal in terms of its fixed carbon content, or by its calorific value: the number of heat units (BTU) produced by a pound of that particular class of coal when free of extraneous mineral matter such as ash. The various ranks of coal are tabulated in table 3.1.

Table 3.1. Ranks of Coal

Type of coal	Approximate Calorific Value (BTU per lb)
Anthracite (hard coal)	14,000
Bituminous (soft/black coal)	12,500
Sub-bituminous (gray coal)	10,000
Lignite (brown coal)	7,000
Peat	3,000

4 The Four Main Classes of Mining Activity

A mine is an excavation made in the earth for the purpose of extracting useful minerals. Such a mine can be established on the surface or underground (see figure 4.1).

UNDERGROUND MINING

An underground mine is not, as the man in the street believes, merely a hole in the ground. If it were, operations would soon have to cease for reasons of safety and economy. On the contrary, to ensure that a mine will produce safely and economically throughout its working life, until the whole mineral deposit has been worked out, it must be developed systematically.

This means the deposit must be sectionalized into local areas that can be conveniently and safely exploited without prejudice to the stability of neighboring areas, and that will enable the ancillary processes of haulage, drainage, and ventilation to be coordinated into a planned, logical layout. The system of development depends upon the configuration of the deposit and on the planned scale of operations. It is different for a vein or lode than it is for a horizontal-type bedded deposit.

The main access opening to any underground mine, whether a vein type or a bedded deposit, is a shaft, where the surface terrain is comparatively level. In mountainous country, access to the deposit may be achieved more effectively by an adit opening instead of a vertical shaft, or by a slope, or inclined shaft.

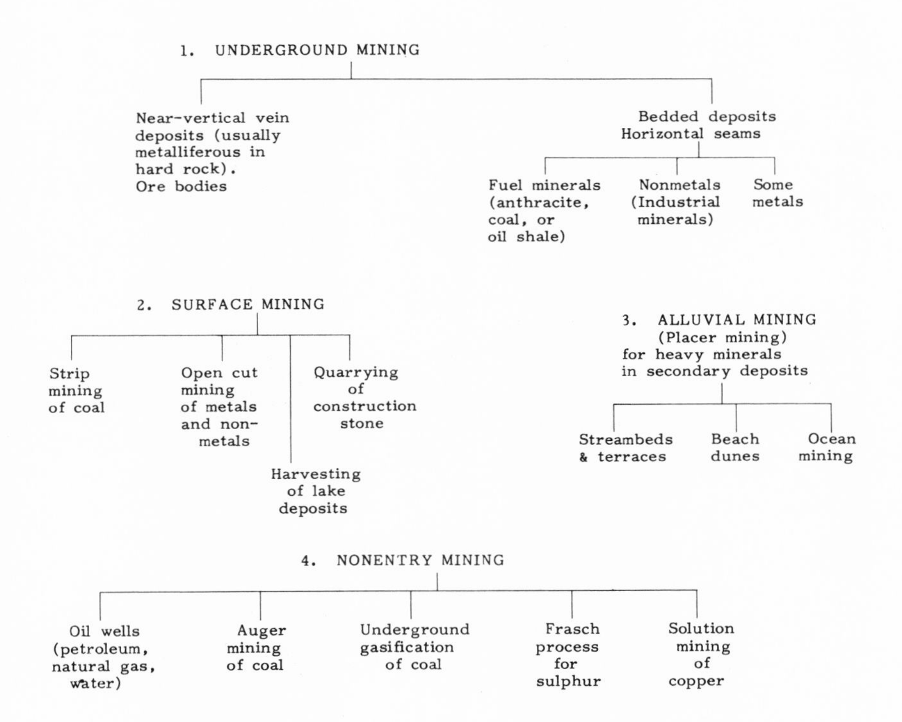

Fig. 4.1. Four classes of mining operations.

Near Vertical Vein Deposits

The shaft is usually sunk in the footwall rock, in order to preserve the stability of the mine throughout its working life (see figure 4.2). Such a shaft is divided into compartments through which particular operations are confined. For instance, a pair of compartments may be set aside as travelling ways through which skips (carrying ore) can be hoisted to the surface. Or alternatively, cages can replace the skips for part of the day for hoisting and lowering workmen, material, and supplies. These cages or skips are hoisted by means of stranded steel wire ropes powered by hoisting engines at the surface, with a headframe erected over the collar of the shaft.

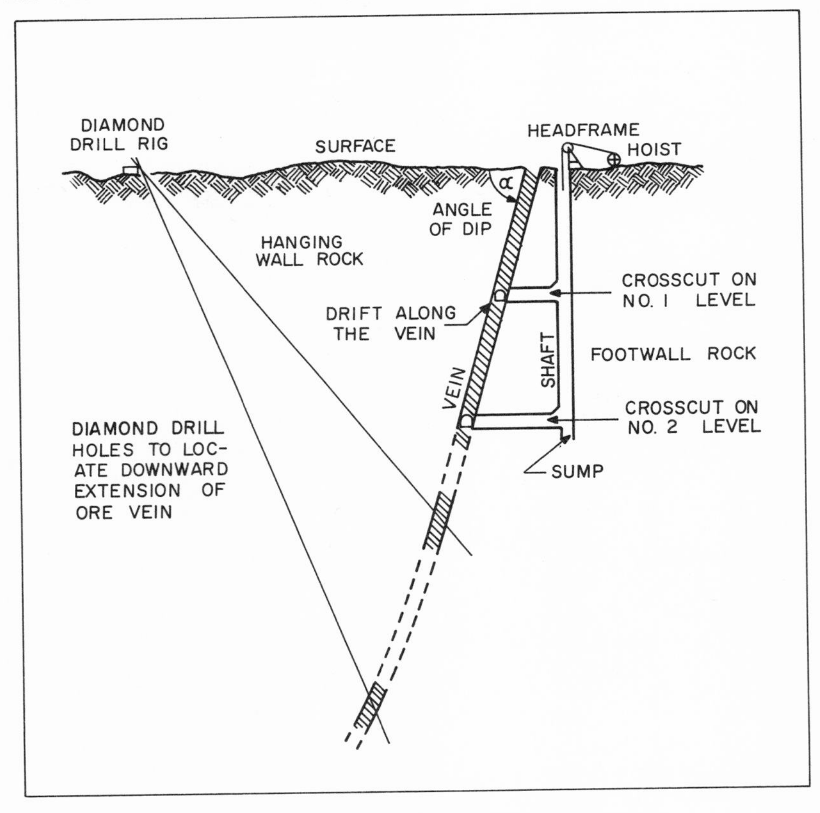

Fig. 4.2. Cross-section of typical ore vein.

Another compartment is designed to be equipped with ladders set on platforms or stages at intervals, and may also carry pipes and electric cables. All compartments provide an airway for the passage of fresh ventilating air into the mine. But this means another shaft is necessary to exhaust foul air from the mine. A large fan is mounted over the collar of this second (ventilation) shaft to provide a sufficient flow of ventilating air through the mine.

For expository purposes, we are going to assume a hypothetical vein of ore 1,200 feet long, with a regular average width of 6 feet, and dipping at about 75 degrees. For such a mine, a crosscut is driven from the main shaft to and through the vein at vertical intervals of about 100 feet. Drifts are then driven along the vein throughout its strike length, which in our hypothetical case is of the order of 1,200 feet.

The workings on each of these separate horizons are called levels, and designated as such in numerical order. At intervals of about 200 feet along each level drift, raises are excavated in the ore of the vein to meet the next level above. By this procedure, the vein is sectionalized into blocks of ore about 200 feet long and 100 feet high (see figure 4.3). These blocks are therefore open or exposed at the top and bottom and at each end. Each such ore block provides a base for the extraction of ore by one of a number of classical methods of stoping.

Following the development of such an underground metalliferous deposit, a particular stoping method must be selected to provide the most economic recovery of the useful mineral, consistent with safety. The main factors to be considered in selecting this method are (1) the physical characteristics of the deposit, in terms of the vein width, the angle of dip, the relative regularity of the deposit and the strength of the ore, (2) the nature of the enclosing rocks, (3) the preservation of the natural environment, and (4) the total costs involved. Naturally, the operation must still yield a profit after subtracting the costs from the revenue derived, in any given period.

The ore block shown in figure 4.3 cannot be removed in toto to the surface. It must be broken into small chunks. The ore is very hard, and therefore it can be broken only by blasting with explosives.

In order to place the explosives within the mass of ore to break it, holes are drilled into the ore with pneumatic rockdrills. These blastholes are charged with explosives and fired at the end of each shift. The resulting dust, heat, and gases are then expelled from the mine by the ventilation system while the miners are changing shift.

The broken ore chunks gravitate through chutes controlled by chute gates to fill ore cars that are hauled in trains along each level to the shaft. Here the ore is transferred into skips and hoisted to the main ore bins on the surface.

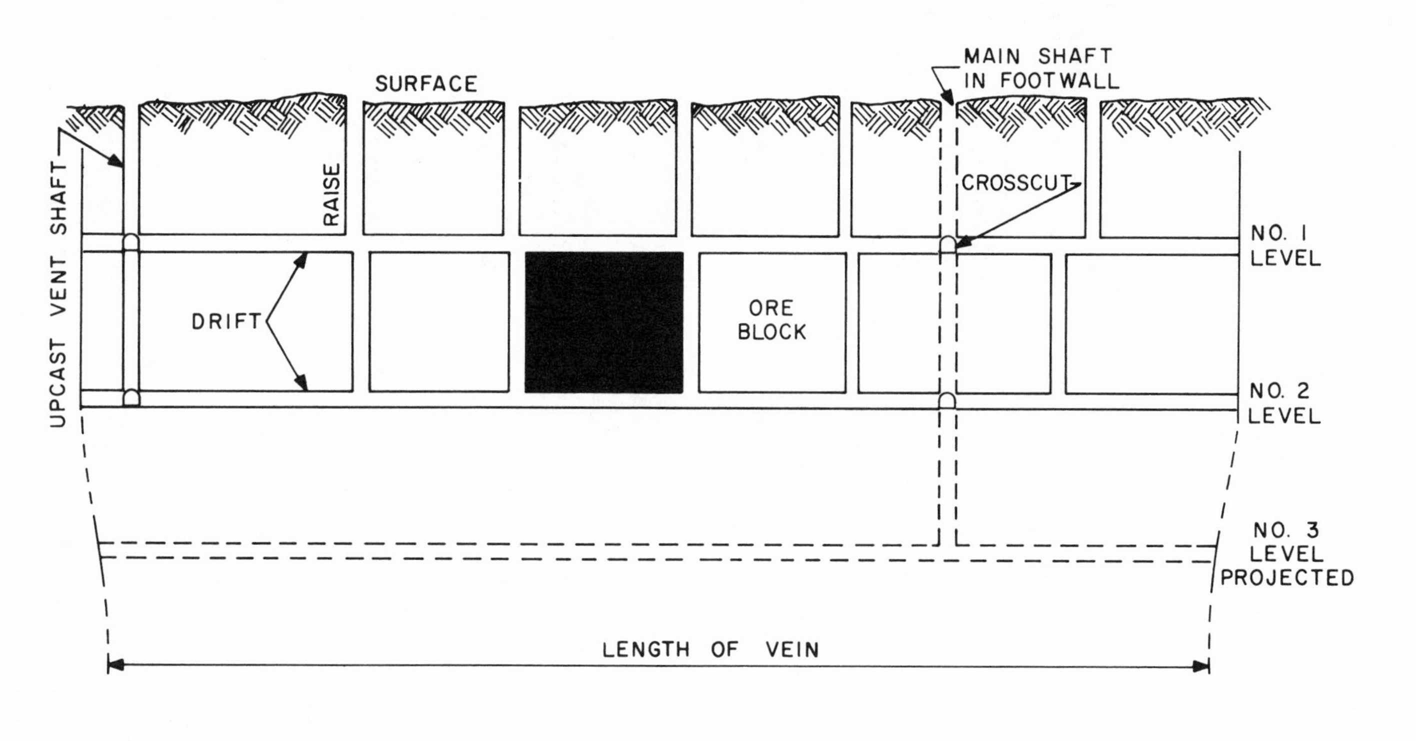

Fig. 4.3. Longitudinal section of ore vein shown in Figure 4.2, with typical development openings on two levels.

One of the classical methods for steep narrow veins where the ore and wall rocks are reasonably strong is known as shrinkage stoping (see figure 4.4). Here, the ore block is prepared by excavating a series of chute raises for a distance of about 12 feet above the lower level. These are connected at the top by a subdrift, and the entrance to each chute is funnelled out to provide an easy passage for the broken ore to pass through the chute without undue blockage. Chute gates are erected at the bottoms of the chute raises to control the flow of broken ore into ore cars.

Stoping is now ready to commence. In our hypothetical case, holes are drilled horizontally from each end raise and a quantity of ore is blasted down, some of which will flow directly into the nearest chute. To prevent the ore from spilling out into the main end raises, a barricade is provided at each end of the stope. This barricade consists of round wooden timber stulls wedged between the hanging walls and foot walls at about five-foot intervals. Behind these stulls, hardwood planks are spiked in a vertical arrangement. As more ore is mined and the back of the stope proceeds slowly upward, more stulls and planks are placed, extending the barricade upward, but still leaving enough room for workmen with their tools to enter the stope, and for ventilating air to pass through.

In this way, slices of ore perhaps eight feet high are stripped off the back of the stope. Sufficient broken ore is left in the stope to provide a working platform for the miners and to support the walls of the excavation. But about one third of the ore is progressively drawn off through the chutes in a controlled manner to allow for the swell factor.

When the back reaches a point about ten feet below the level above, the stope is completed, leaving a crown pillar above to protect the upper level. At this point, all the remaining ore can now be drawn out through the chutes, and the resulting stope cavity can, if necessary, be filled with waste material to prevent collapse of the walls. At a later stage of the life of the mine, the crown pillar and the chute pillars can usually be mined, by a special program which we will not consider here.

This is a simplistic description of a standard shrinkage stope. All sorts of variations have been adopted in different mining districts. Another classic method of mining is known as cut-and-fill stoping, as shown in figure 4.5. Here, the ore block may be prepared in the same way, but the chute raises are not funnelled out at the top. After every eight-foot slice of ore has been stripped from the back, a series of specially cut sets of standard timbers are built up above each chute to within about six feet of the back. Waste filling material is now placed in the stope between adjacent timbered chutes and betwen the end chutes and the barricades. Several alternative

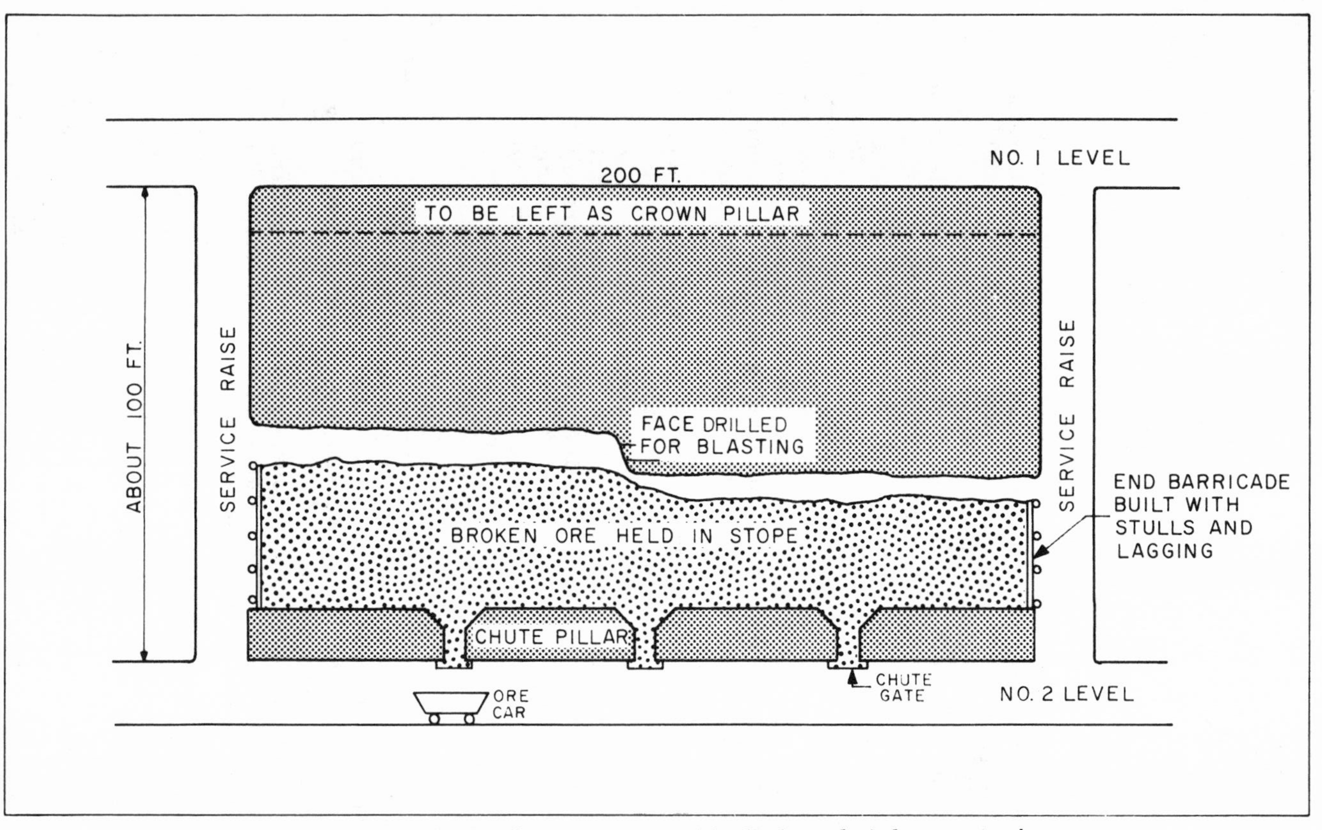

Fig. 4.4. Method of mining ore block by shrinkage stoping.

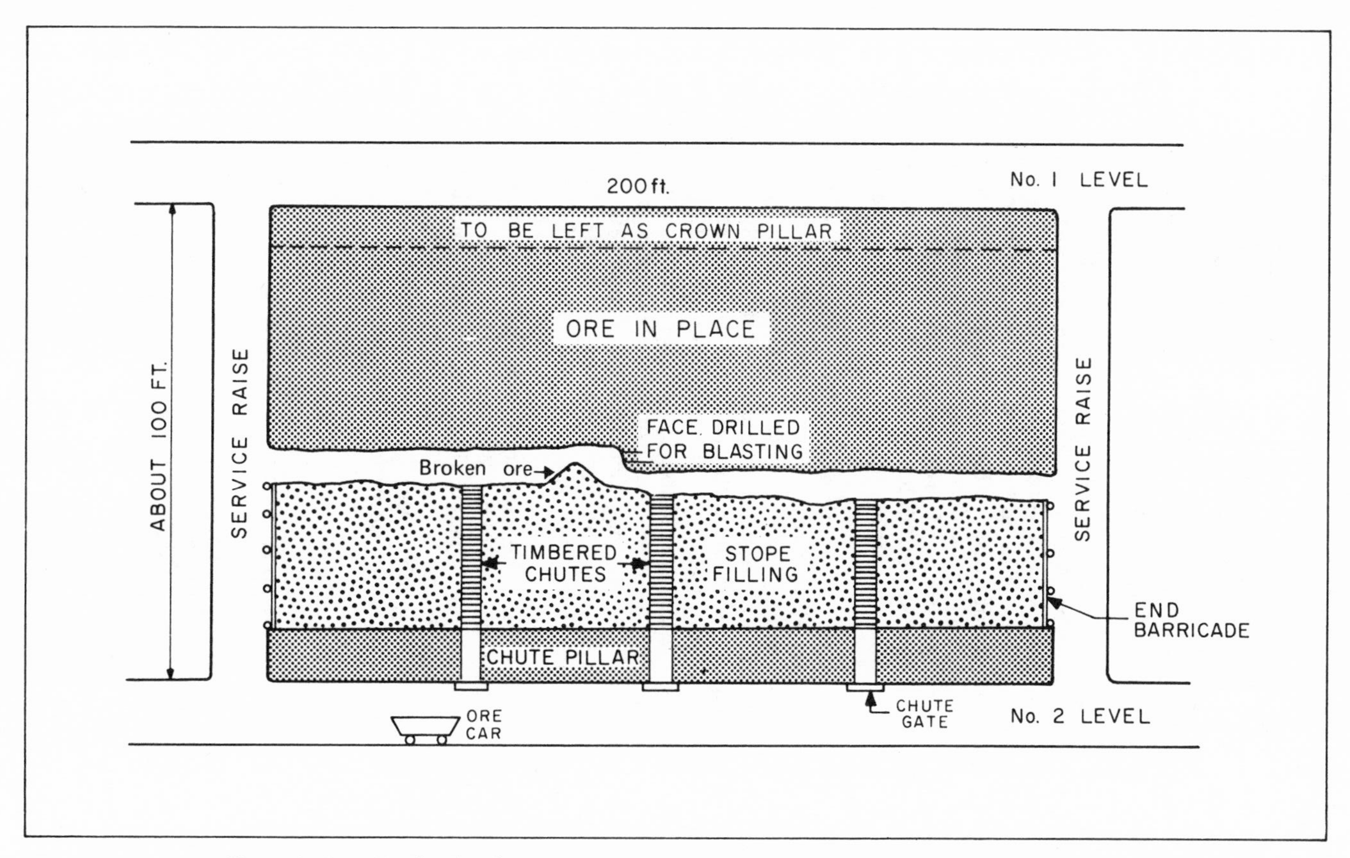

Fig. 4.5. Method of mining ore block by cut-and-fill stoping.

filling materials and methods may be used, but we will not enlarge on them here.

The top of each layer of filling now provides the working platform and the broken ore is loaded or scraped into the chutes. As the stope proceeds upward, timbering and filling proceed on a cyclical basis. When the crown pillar is reached, the stope is completed and abandoned. Various alternative types of cut-and-fill stoping are found in different mining districts.

There are several other standard methods of stoping narrow veins but the few just discussed are sufficient to give an insight into the general stoping procedures.

Of course, for wide lodes and other configurations of ore bodies, a number of other stoping methods are available, the choice depending upon which one more closely applies to the physical characteristics of the ore and walls. In many of these cases, there is sufficient room to deploy mobile mechanized drilling and loading equipment under mass-production conditions.

Year by year, as mining is being carried out at successively deeper levels, more and more problems arise. For instance, much more water seeps into the mine workings and must be pumped out against higher heads. The ore must be hoisted a greater vertical distance, and therefore it takes more hours per day to hoist a given tonnage. Rock pressures become greater and the stability of the mine, now becoming more and more honeycombed with old workings, becomes a serious matter. Rock temperatures also increase with depth, and this, together with the increased airflow resistance offered by a deeper mine, calls for greater quantities of ventilating air to be circulated.

Some mines in the world, particularly those in South Africa, are very deep. The Western Deep Levels Mine is now down over 12,600 feet (more than two miles) vertically, and is planned to be mined to a depth of 13,000 feet. In mines like this, the ventilating air must be refrigerated (air-conditioned) so that comfortable conditions may be provided for the working miners.

Horizontal Bedded Deposits

By far the most important types of bedded deposits are coal seams. Hypothetically, we are going to assume that our coal seam is six feet thick and extends in all directions horizontally to the boundaries of the mining property. The sedimentary rock above the seam is usually of shale or sandstone and is termed roof rock. Similarly, the rock below the seam is called floor rock (see figure 4.6).

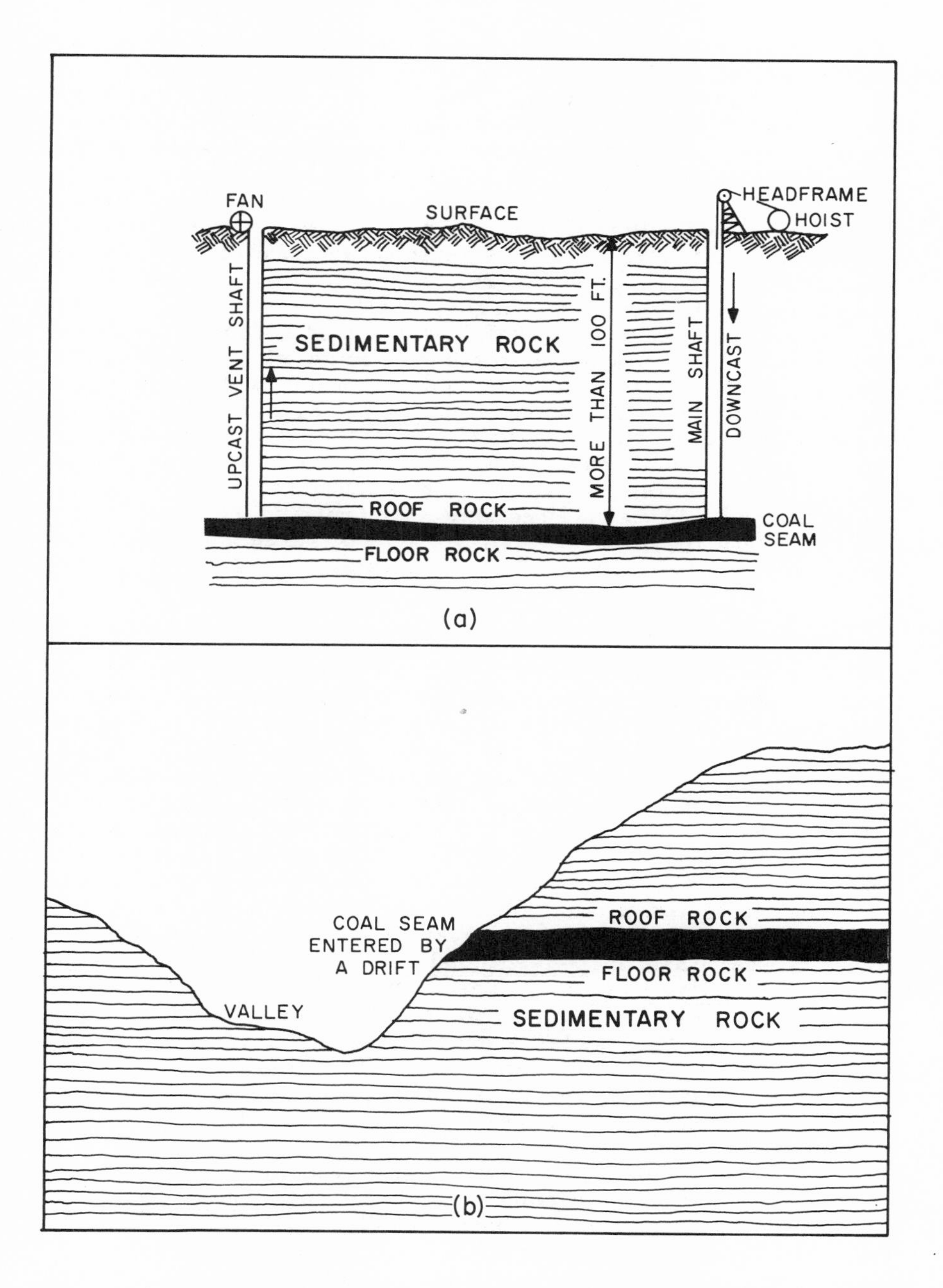

Fig. 4.6. Typical underground coal mine development: a) shaft mine in flat country; b) drift mine in hilly country.

If our seam occurred in mountainous country, then we could gain access to the seam by an adit, in which case it would be called a drift mine. But if, as we assume, the surface terrain is roughly horizontal, we will need to sink two shafts from the surface through the seam. One of these would be a main working shaft and the other a ventilation shaft.

Where the shaft passes vertically through the coal seam, we leave a shaft pillar of unmined coal to protect the shaft and to maintain stability. For this reason, we sometimes sink both shafts in the same general area so as to use the same, although perhaps a somewhat larger, shaft pillar.

Apart from these two shafts which pass through the overlying sedimentary rock, all other workings will be within the seam of coal, so that all excavations made will incidentally produce coal. In order to develop the mine, our huge slab of coal, in this case six feet thick and extending horizontally to the property boundaries, will be dissected into sections, or "districts" by extending main entries from the shaft pillar to the boundaries in each direction. The mine is now ready for the systematic removal of coal, by one of three main methods: the room-and-pillar, longwall, or shortwall method.

Room-and-pillar method

In the room-and-pillar method the coal is extracted by driving "rooms" about 18 to 20 feet wide on about 60-foot centers. This means that a pillar of coal about 40 feet wide is left intact between adjacent rooms. After proceeding in the rooms for about 60 or 80 feet, crosscuts are made at right angles to the general direction of mining. These crosscuts link up the rooms, and we get a rectangular pattern of unmined pillars about 40 by 60 feet to support the roof rock (see figure 4.7). Meanwhile, all the work performed in driving rooms and crosscuts results in the production of coal.

These rooms and crosscuts are advanced by two alternative systems: (a) the conventional mining system, and (b) the continuous mining system. The conventional mining system uses a slot called a kerf which is cut across the face of the coal at floor level. Holes are drilled above this and charged with special "permissible" explosives. When the holes are fired, the broken coal is loaded by a loading machine into transport vehicles to feed a belt conveyor in the main entry for haulage to the shaft.

The continuous mining system involves the use of a highly integrated mobile machine called a continuous miner (see figure 4.8). This machine mechanically rips coal out of the face as it keeps moving forward. The broken coal falls onto a steel apron that feeds a short inbuilt conveyor for transferring the coal to the rear of the machine. Here the coal is handled by one of a variety of methods to deliver it to the main con-

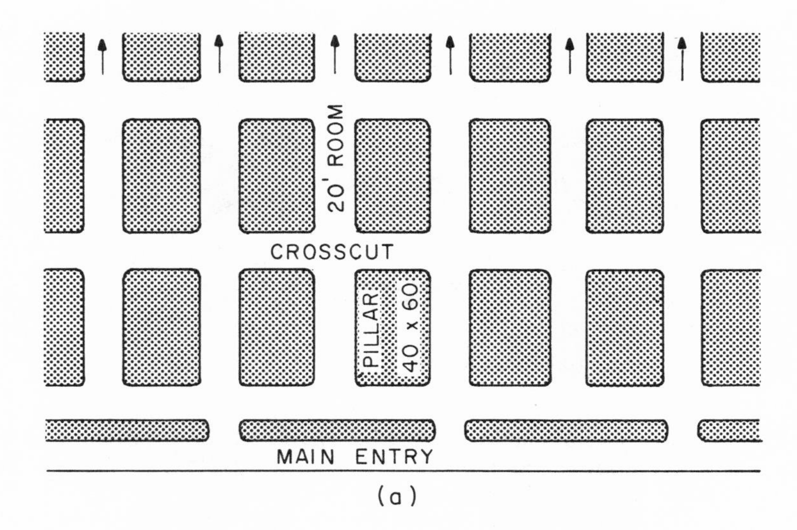

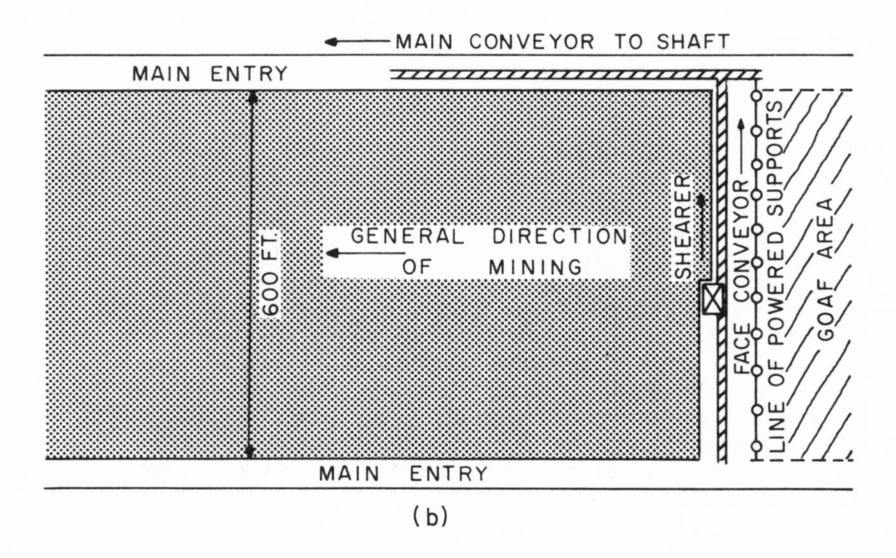

Fig. 4.7. Two methods of mining coal underground: a) room-and-pillar method and b) longwall method.

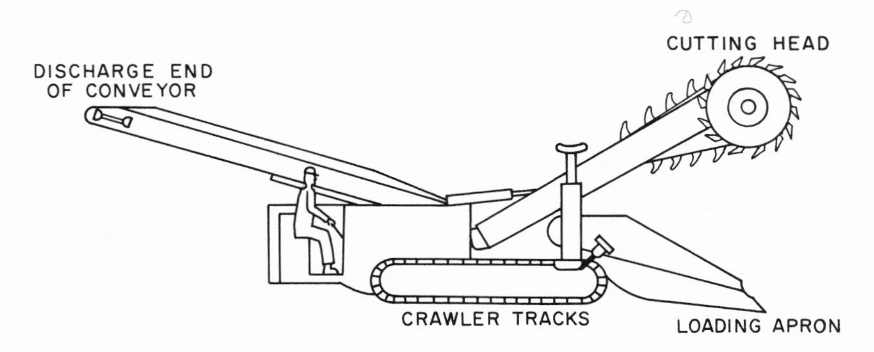

Fig. 4.8. A continuous miner.

veyor system in the main entry. With either system, particular attention must be paid to the need to allay the dust, to ventilate the faces, to remove seepage water, and to give additional support to the roof, either by wooden props or roofbolts.

One of the chief disadvantages of the room-and-pillar method is that most of the coal is left behind in the pillars. This would represent a national waste, but most companies attempt to recover the pillar coal by special methods at a later stage of mining. At least some of the pillar coal is recovered in this way.

Longwall method

The longwall method has many advantages over the room-and-pillar method, chiefly because very little coal is left behind unmined in the pillars. Large panels of coal are initially developed perhaps 600 to 1,000 feet in width (see figure 4.7). In retreating from the boundary, the whole face is mined in slices, back toward the shaft. A cutting machine skims coal from the face and dumps it on a face conveyor belt which then delivers it to the permanent conveyor in the main entry. To support the roof near this long face, a line of specially designed cantilevered powered supports, like massive hydraulic jacks, is maintained in position. As the face advances, the face conveyor and the powered supports are moved forward in a regulated program. Meanwhile the roof is allowed to collapse behind the line of supports in what is known as the gob, or goaf area.

Shortwall method

The shortwall method is a hybrid method developed in Australia. It involves the better features of the room-and-pillar and the longwall methods, in that a continuous miner operates under the protection of the type of cantilever roof support used in the longwall method.

When the softer types of industrial minerals, such as potash and trona, occur in bedded deposits, they are mined in the same general way as coal seams, using continuous miners in a room-and-pillar method, or otherwise, in the longwall method. With hard industrial minerals and with metalliferous ores in bedded deposits, the face between the pillars needs to be drilled and blasted. In these cases, pillars are left in an irregular pattern, mainly only where (a) the roof rock is locally weak, or (b) the ore is of submarginal grade.

SURFACE MINING

Where a bedded-type mineral deposit occurs at or near the surface, or where a wide lode exists with a large surface expression, it can be mined at a lower cost and with fewer problems by surface mining (see figure 4.1). Coal seams lying within 100 feet of a reasonably level surface can be conveniently mined in this way. This is known as strip mining, because the superincumbent overburden is first stripped off and cast aside in order to expose the seam of coal in strips (see figure 4.9). After the coal is removed, the overburden is replaced and the area is revegetated.

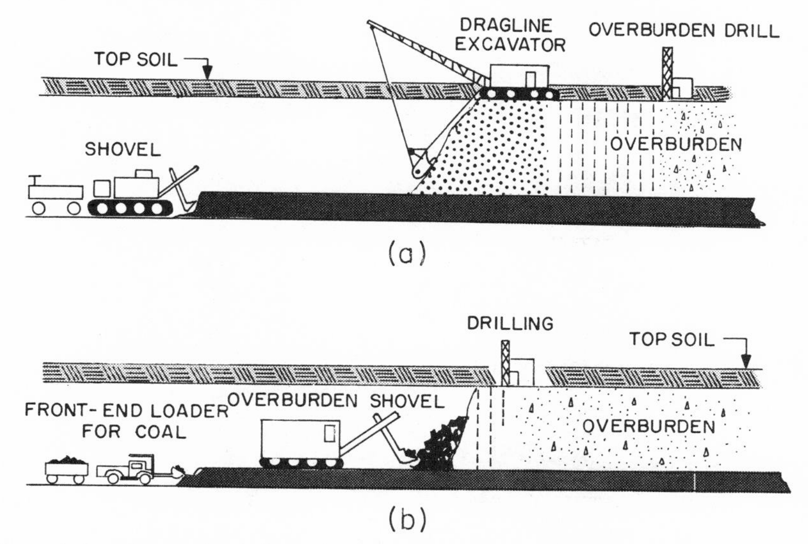

Fig. 4.9. A coal strip mine operation: a) by dragline excavator; b) by power shovel.

A slightly different procedure is involved in mountainous country where the coal seam outcrops around a particular contour of a mountain. This is called contour strip mining. As much of the overburden as possible is removed and stacked in a valley to expose coal until the resulting highwall is likely to become dangerous. After mining the exposed coal around this contour, a further quantity of coal can be recovered from beneath the mountain by means of auger mining.

Strip mining is usually carried out on a large scale. For this purpose, gigantic machines like walking dragline excavators, power shovels, or bucket-wheel excavators are used to strip and dump the overburden (see figures 4.10-12). The

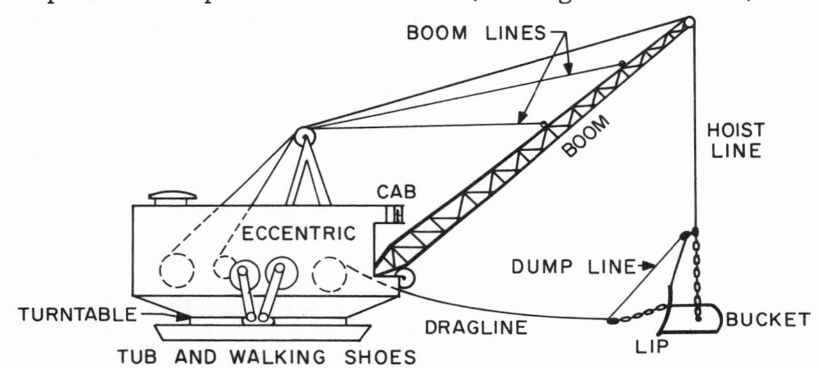

Fig. 4.10. A walking dragline excavator.

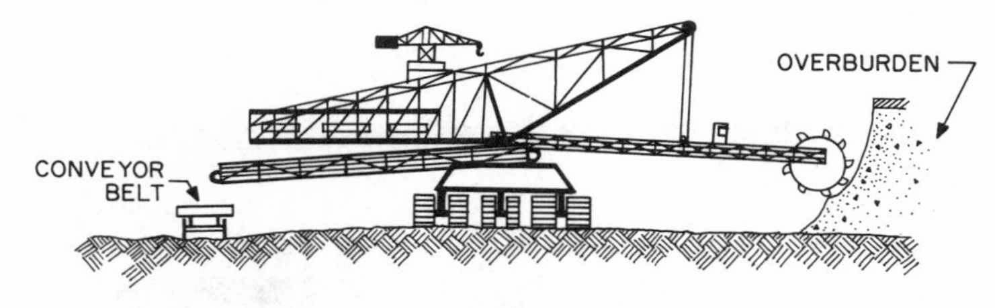

Fig. 4.11. Bucket-wheel excavator.

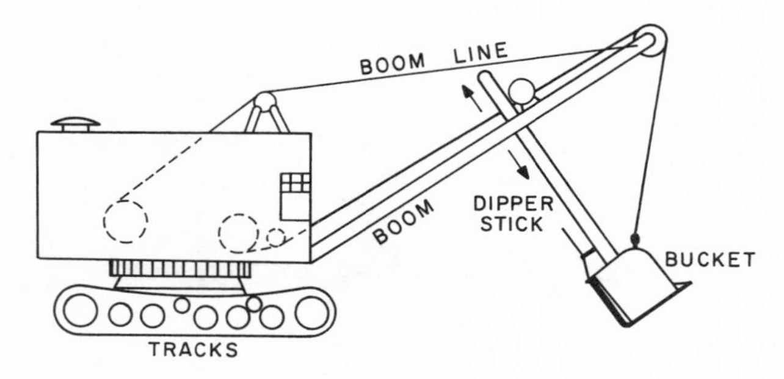

Fig. 4.12. A power shovel.

coal is then mined with smaller power shovels, loading into off-highway trucks. In most cases, a certain amount of drilling and blasting is required to loosen hard bands of overburden material, and therefore to facilitate the action of the excavators. Bedded deposits of the softer industrial minerals, occurring near the surface, can be handled in the same general way.

The harder industrial minerals and the metalliferous disseminated or replacement deposits may also be handled by surface mining, generally called open cut work. These deposits are often no more than 40 feet thick, under overburden of a maximum depth of say, 50 feet. In these cases we are dealing with hard, rocky overburden and ore. Much more sophisticated methods of drilling and blasting must be used, and a positive type of loading machine like a power shovel is called for. Many of the porphyry copper mines in Arizona and New Mexico have adopted these procedures.

For steeply dipping wide lodes with a considerable vertical extent, additional considerations apply. After preliminary overburden removal to expose the whole of the ore body, a series of "benches" is developed. These are stepped in heights limited to about 50 feet for safety reasons. The horizontal part of the bench is designed to provide space to accommodate the broken ore from each blast, to deploy drilling machines to drill blastholes for blasting on the next lower bench, to allow space for mobile power shovels to move in and load the ore into transport vehicles, and to provide a roadway for the latter. The general setup is shown in figure 4.13.

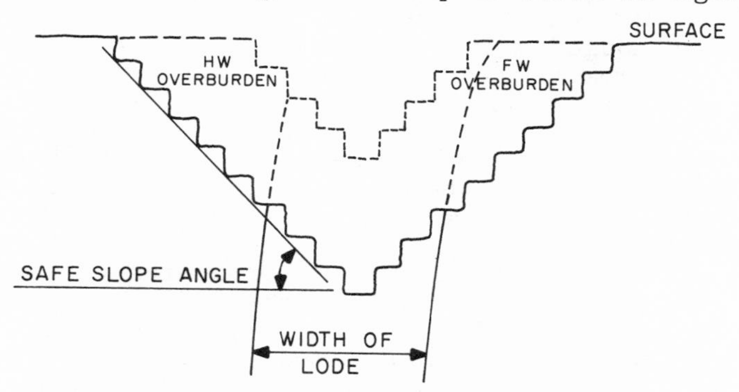

Fig. 4.13. Open cut mining of a wide lode.

However, as open cut mining proceeds to greater depths, the exposed hanging wall and the footwall would become a source of hazard unless these walls were stripped back (by a similar benching procedure) to provide a safe general slope angle to the bottom of the cut. The waste rock mined in the walls is transported to a dump area, whereas the ore is taken to the crushing plant and the mill.

As mining proceeds to deeper horizons, increasing tonnages of overburden are mined from the walls, adding considerably to working costs. A point is inevitably reached where it is better to cease open cut work and to continue to mine the deposit by converting it to an underground mining operation.

Quarrying is another type of surface mining operation, where rock is being mined (or quarried) for the production of construction stone or building stone. Similar methods apply to those for open cut work in metalliferous deposits.

Another phase of surface mining relates to the harvesting of salt and other minerals from dry salt lakes or from lake brines as at Searle's Lake, California, and the Great Salt Lake in Utah. A number of minerals corresponding to sodium, potassium, and magnesium chlorides and sulphates are produced by solar evaporation, fractional crystallization, and by mechanical scraping and loading of the respective materials.

ALLUVIAL MINING

Alluvial mining is sometimes called placer mining. It relates to the recovery of heavy (high specific gravity) minerals that have become concentrated in secondary deposits, following their separation by weathering agencies from primary veins higher up the watershed.

These dislodged minerals eventually become concentrated, typically in streambeds, below the gravels and immediately above the bedrock. Many collect in depressions in the top surface of the bedrock. Only the heaviest of minerals are concentrated in this way. Others are carried along the streams into lakes or to the oceans, where they become concentrated in a different way by wave action, in beach sand or dune deposits.

The minerals more susceptible to concentration in alluvial deposits range from platinum (which when pure has a specific gravity of 21.4), and gold (19.3), through cassiterite (tin oxide), diamond, garnet, monazite, magnetite, zircon, to rutile (4.2). By contrast, the specific gravity of ordinary rock minerals, sand, and gravel is 2.6. Some alluvial deposits are found in present streambeds. Others are found in terraces and in deep buried areas which were the beds of former streams. For this reason, although most placer-mining activities are carried out as a surface operation, others (such as deep leads) are mined by shallow underground methods.

Deposits like those just described are readily found and easily mined by relatively simple methods. Therefore, most of the world's placer deposits (except those concealed by snow and muskeg in Siberia) have already been found and worked out during the last 5,000 years or so.

The rich, high-grade gold placers were mined simply by panning the gravels, or by shovelling the gravel into cradles or sluice boxes to shed the lighter material and to collect the gold behind riffles and similar traps. Mercury is sometimes used to collect fine particles of gold and silver by a process of amalgamation. The gold is later separated by retorting the amalgam, the mercury becoming vaporized and later saved by a condensing apparatus; the resulting sponge gold is melted into bars.

Where gold-bearing gravels occur in the banks or terraces of streams, they can often be mined by sluicing away the bank material with a hydraulic monitor, using high-pressure water. The gravels are then washed through large sluice boxes to collect the gold which settles behind the riffles.

However, fine gold may travel along streams for many miles before it reaches a wide river valley. Here the original stream loses much of its velocity and the fine gold is deposited in the slower moving stream in widespread gravel deposits. The gold content of these gravels is generally very low, but large-scale dredging methods can often recover it at a profit.

Tin ore (cassiterite) is similarly recovered from river valleys by large integrated dredging units in Malaysia, Thailand, and Indonesia (see figure 4.14). Heavy mineral sands found in beach strands, or in dunes above the beach, are also mined by hydraulic monitors or by dredging. Chief among these deposits are the beach sands of eastern Australia from which rutile, ilmenite, zircon, and monazite are recovered. These heavy beach sands are then treated in concentration plants to recover the minerals separately. The lighter clean silica sand is returned to the beaches (see figure 4.15).

Ocean mining relates to a new branch of the mining industry, recently under development to recover nodules of manganese dioxide from the sea floor, at depths ranging from 100 to 15,000 feet. The nodules are reported to contain metal values of manganese, copper, nickel, zinc, cobalt, and molybdenum. Various methods such as dredging are currently being devised to recover these nodules and to deliver them to ocean-going barges for transfer to treatment plants on the shore.

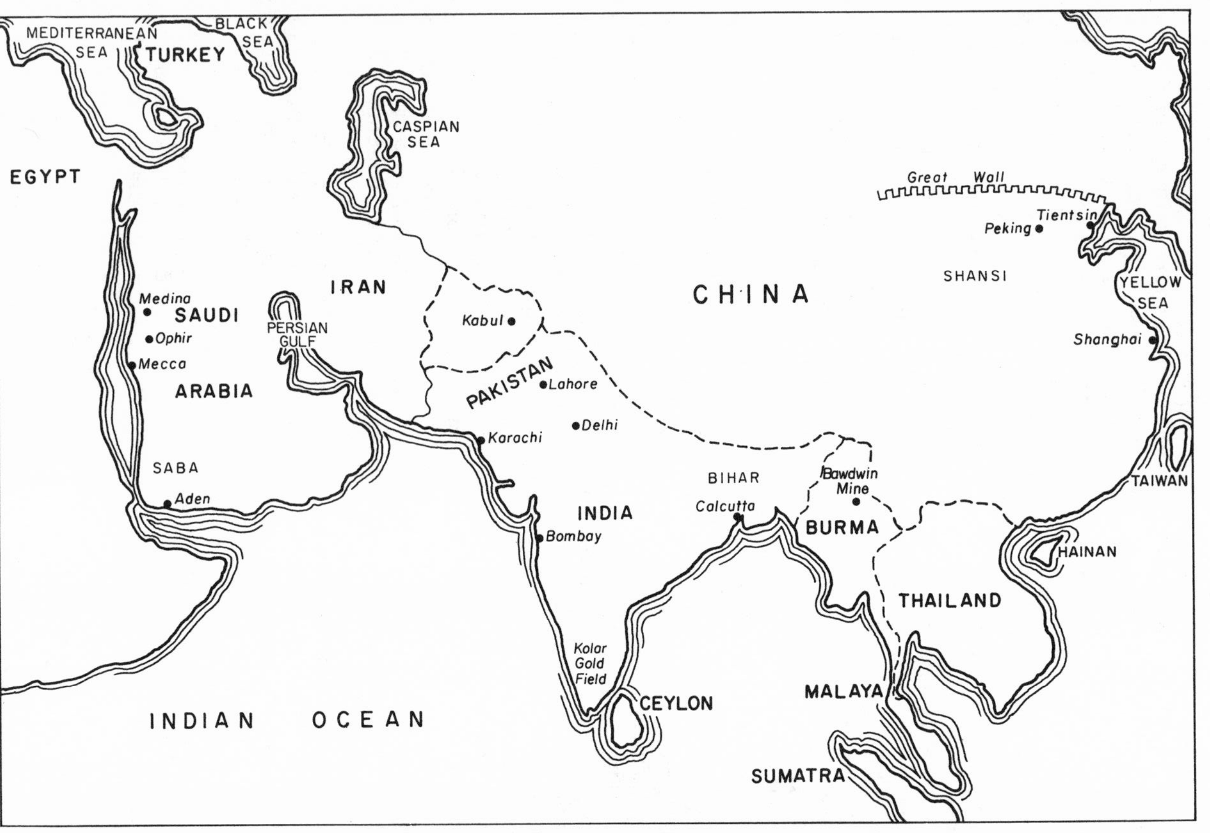

Fig. 4.14. Map of Southern Asia.

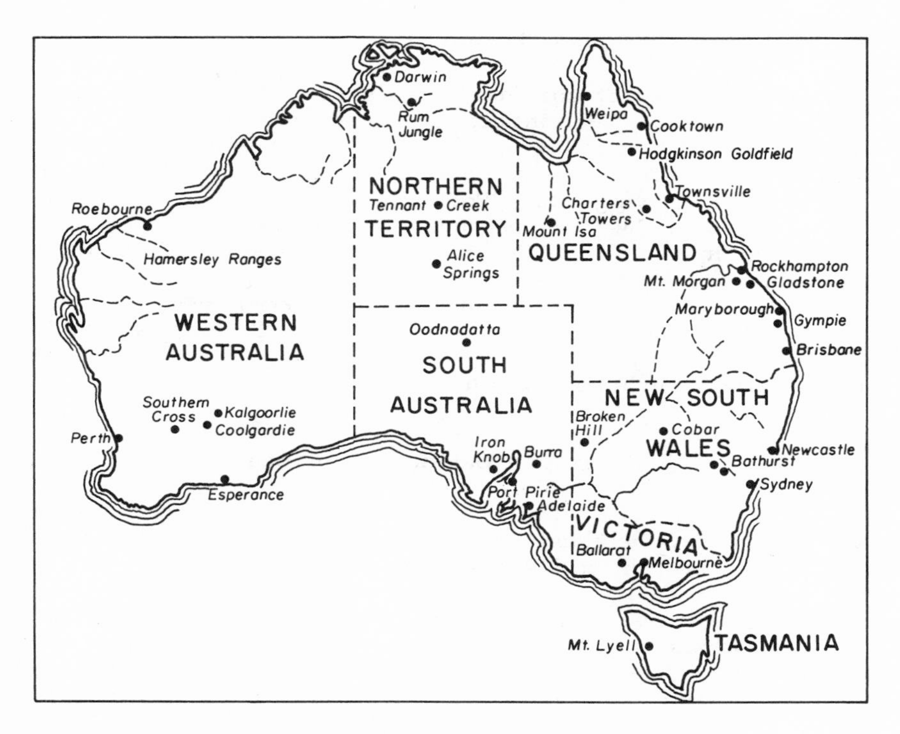

Fig. 4.15. Map of Australia

NON-ENTRY MINING

Under the heading non-entry mining we have a number of widely differing approaches for recovering deep-seated minerals without mining them by conventional underground methods. This simply means that miners do not have to go underground to bring these minerals to the surface.

Oil Wells

Crude petroleum and natural gas are brought to the surface by drilling oil "wells" to depths ranging up to 20,000 feet to tap deep-seated oil deposits. The oil is usually brought to the surface by pumping, aided by applied pressure. Oil well drilling rigs are highly specialized units, using tri-cone roller-type noncoring bits in diameters up to 12 inches.

Auger Mining of Coal

Auger mining of coal has already been explained. These large augers bore horizontally into coal seams exposed in the side of mountains, following the contour-stripping operation. They are built in diameters up to 6 feet and are capable of drilling and removing coal (like a giant carpenter's auger bit) from a depth of 200 feet measured horizontally. The coal, as it emerges in the flights of the auger, is elevated into a bin for loading transport vehicles.

This is a remote procedure. No workmen need to enter the area from which the coal is extracted. It enables coal to be recovered from under and behind the highwall without the need to strip overburden further.

Underground Gasification of Coal

Extensive trials have been made in Britain, the United States, France, Belgium, Poland, Italy, and Russia to gasify poor quality or thin coal seams in situ underground. It avoids the need for men to work underground, especially in these thin inaccessible seams. The principle is relatively simple but the applications in practice are difficult. The main components of the system are: (a) an inlet shaft or borehole, to admit compressed air; (b) a gasification area where the seam is fractured by a jet of compressed air, and is ignited; and (c) an outlet shaft or borehole to convey volatile gases to the surface. Normally, a series of boreholes is needed to extend the combustion zone once operation is commenced.

One of the main problems is that, so far, maximum efficiency is not obtained because the system yields volatiles only. Nevertheless, it has other important advantages.

The Frasch Process for Sulphur Recovery

Sulphur-bearing strata are associated with buried salt domes in the Texas gulf area. To avoid mining these deposits by conventional underground methods, Dr. Frasch invented a non-entry method in 1894.

Boreholes are drilled from the surface into the sulphur-bearing strata. Superheated water or steam is pumped down into these strata to melt the sulphur. The molten sulphur is forced to the surface in a central pipe by hot compressed air. This method produces relatively pure sulphur.

Solution Mining

In the old salt mines of Austria, water flows into the rock salt deposits, dissolves the salt, and is pumped back as a brine solution to the surface for evaporation of the water and precipitation of the salt.

Underground deposits of copper or uranium ore may be leached in situ by percolating leach solutions through them, collecting the pregnant solution, and pumping it to the surface for precipitation of the copper or uranium salt. The successful operation of this method depends upon the permeability of the ore body. Various methods of prefracturing the ore body are under investigation.

5 The Scope of Mining Activity

It is important to understand the great scope and complexity involved in the various stages of establishing and operating a mine from the grass roots until the ore deposit is completely mined out. Naturally, these activities are much more complex in the case of the near-vertical vein and lode deposits (which may call for the extension of the mine workings to great depths in extremely hard rock) than they are for the horizontal bedded deposits so typical of coal seams and beds of industrial minerals; and they are certainly much more complex than any type of surface mining. To cover the whole range of these activities, it is, therefore, considered necessary to describe the stages involved in the exploitation of such a metalliferous vein or lode type of deposit.

In the preliminary stage, by various techniques of prospecting and exploration, we locate a mineral deposit and stake the ground, or acquire the deposit under an option agreement or by direct purchase. Then we need to employ costly and sophisticated methods of diamond drilling and further underground exploration work to evaluate the form, shape, grade, and tonnage available in the deposit.

The next step is to prepare a feasibility report to provide basic evidence of the probability of mining the ore body over a period of years that will yield a profit commensurate with the risks involved. Such a feasibility report needs to be extremely comprehensive. It should incorporate the study of alternative mining methods with the associated degree of development planning, the projected working cost estimates of current development work; of stoping the ore, of crushing, grinding, and concentrating the ore; of transport of the concentrate to a smelter and the smelting costs, and of transport of basic metal products to a refinery or otherwise to market. All of these estimates need to be prepared in terms of detailed labor needs and costs, work efficiencies, costs of supplies, maintenance, power, water, and overheads. The overall working costs must then be related to the projected overall earnings year by year from sale of metal products based upon projected metal prices (which are customarily quite variable), or upon long-term sales contracts that may be anticipated. In this way, a cash-flow table and an overall profit estimate can be determined.

Armed with this feasibility report, prepared and checked by competent mining engineers, the board of directors can then make a decision whether or not to form a company and proceed with the opening and operation of a mine. If they make an affirmative decision, the development stage comes next. This is necessary to prepare the deposit for mining by a well-planned system of mine development, including the sinking and equipping of a shaft. Similarly, the metallurgical testing of a representative sample of the ore is carried out. Based upon the results secured from these tests, the design of a concentrating mill is prepared, materials and equipment are ordered, and erection of the concentrator proceeds. Similar procedures are used if the erection of a smelting plant is also involved.

During the same period, and especially where there are no nearby towns or facilities adjacent to the mine site, an ample infrastructure program would need to be provided. This could include the provision of employee housing, workshops, powerhouse, water supply, amenities, health services, roads, railway, airport, and perhaps even a seaport.

Depending upon the scale of operations proposed, these first two stages of establishment could take up to five years of lead time and call for the capital expenditure of hundreds of millions of dollars - before a single ton of metal can be produced or sold, or a single dollar earned.

The third stage is the production stage, during which actual mining of the deposit begins by using a particular method of stoping, mainly using drilling, blasting, and loading techniques in each stope. Associated activities include arrangements for sampling and surveying, pumping the seepage water out of the mine, providing adequate ventilation and lighting and other facilities, supporting the various excavations to prevent collapse, hauling the broken ore in ore trains along the various levels from the stopes to the shaft, hoisting the broken ore up the shaft to the surface and dumping it into ore bins, as well as a wide range of detailed minor activities.

From the surface ore bins, the ore is drawn out and fed successively to crushers and grinding mills. The ore must be ground into extremely fine particles to separate the valuable minerals from the worthless rock materials by some method of concentration. The worthless gangue materials are discarded as tailings, most of which are sent back underground to fill the empty stopes for support of the walls.

The mineral concentrate is then filtered and sent to the smelting plant where fluxing materials are added and the whole charge is roasted and smelted to produce (a) a worthless slag, and (b) ingots of the metal which can then be sold. Or alternatively, these ingots can be sent to a refinery where a more refined metal and perhaps several co-product metals are produced for the market.

The time involved from the breaking of the equivalent of one ton of metal as ore in the stope to the receipt of the metal ingot by the buyer as a basis of sale may be of the order of six months. Consequently, a considerable amount of working capital is required.

Production then keeps going at a steady designed rate. Mine development work must keep ahead (perhaps 2 years ahead) of ore production activities. As the ore body becomes exhausted, more dollars are called for to find possible extensions of the ore body to provide more tonnage, or alternatively, to locate a new ore body elsewhere. The period during which an ore body becomes exhausted is equivalent to a planned extraction rate that yields a projected life of 20 to 25 years for an individual mine.

In the final stage, when the ore body becomes exhausted, mining and concentrating activities cease, the mine fills with water and is abandoned. Meanwhile, all equipment installed underground is left and written off, and very little of the surface plant is salable above scrap value, especially if the mine is in a remote locality. The restoration and landscaping of the surface dumps, which had been proceeding actively during the mining operations, must now be completed and the various mine entrances made safe. The company is liquidated unless it can acquire another ore body elsewhere.

This simplified analysis will serve to demonstrate the great complexity of operations involved in the establishment and operation of an underground, multilevel metalliferous mine. As the mine workings become deeper with the process of time, the problems involved become significantly greater. There is probably no more complex operation in the whole gamut of industrial activity than that of a multilevel, underground metal mine.

If the mine happens to be near a town site with existing housing, power, water supply, communications, and other facilities, then the infrastructure needs are greatly reduced. Coal mines and others based upon bedded deposits generally involve fewer and simpler procedures.

Following a study of the foregoing chapters the reader (as a neophyte miner) should have gained a basic across-the-board appreciation of the nature of minerals, their mode of occurrence, the various methods of mining and beneficiation of each main type of deposit, and the general scope of the activities involved.

Armed with this background material, he should now be in an enhanced position to study the history of mining with greater zest and appreciation.

II

The Eight Ages of Man

An Introduction to Part II

Archaeology is the science of the evolutionary development of mankind. The first evidences of human existence have been found during the last few years by the celebrated archeologist, the late Dr. Louis S. B. Leakey and his wife and son, in East Africa. Nevertheless, the early discoveries in Africa were made by Dr. R. A. Dart, professor of anatomy in the University of Witwatersrand.

The fossil remains of man are evident only in the bony structures because the softer parts soon become decomposed and disintegrated. The gradual development of man is therefore more closely followed by studies of the artifacts associated with his manner of living. These artifacts include tools, weapons, implements, utensils, and ornaments made from various materials throughout the evolutionary period of mankind. The age of the artifacts fashioned and used by primitive man can now be more precisely determined by the radiocarbon dating process.

It has become convenient to designate various periods of mankind's evolutionary development as the Ages of Man, as represented by his use of various minerals or metals. These include the Stone Age, the Copper Age, the Bronze Age, the Iron Age, and so on (see table I.1). Part 2 covers these developments in a general way (see table I.2). More detailed aspects of developments in mineral technology over the centuries are covered in part 3.

Table I.1. The Eight Ages of Man

500,000 B.C.:	Paleolithic (Old) Stone Age began (using eoliths until 100,000 B.C., when first reports of actual mining were noted).
8000 B.C.:	Neolithic (New) Stone Age began.
5000 B.C.:	Copper Age began.
3000 B.C.:	Bronze Age began.
1400 B.C.:	Iron Age began.
1600 A.D.:	Coal Age began.
1850 A.D.:	Petroleum Age began.
1950 A.D.:	Uranium Age began.

Table I.2. Ages of Selected Civilizations

Sumerian	3500-1900 B.C.
Egyptian	3300-300 B.C.
Cretan (Minoan)	2300-1400 B.C.
Hittite	2100-1200 B.C.
Hebrew	2000 B.C.-135 A.D.
Babylonian	1900-729 B.C.
Aegean (generally)	1800-1150 B.C.
Assyrian	1700-605 B.C.
Phoenician	1200-586 B.C.
Greek (Homeric)	1000-146 B.C.
Carthaginian	814-149 B.C.
Roman	750 B.C.-450 A.D.
Lydian	670-547 B.C.
Chaldean	605-538 B.C.
Byzantine	320-1453 A.D.
Mohammedan (began)	570 A.D.
Carolingian Renaissance (began)	770 A.D.
Middle Ages (medieval times)	450-1450 A.D.
Dark Ages (early Middle Ages)	450-1050 A.D.
Feudal and Manorial System	800-1250 A.D.
Wars of the Crusades (began)	1096 A.D.
Renaissance	1350-1550 A.D.
Scientific Revolution	1550-1700 A.D.
Industrial Revolution (began)	1770 A.D.

6 Paleolithic (Old) Stone Age

The early apelike creatures believed to be the ancestors of true man have been found mainly in Africa, dating back some three or four million years. They are now known as Zinjanthropus, the forebears of the Australopithecus race.

Eight successive waves of these primitive races existed in interglacial periods between various ice ages. These were the forerunners of modern man. They were basically flesh eaters (carnivores) and, therefore, they depended upon wild animals for their food and, in later periods, for clothing. They were essentially nomads, moving from place to place seeking fresh supplies of game for their existence. But they had one special need: a source of good material for fashioning weapons and hunting tools.

Those who lived in western Europe were fortunate because they found excellent deposits of flint stones occurring in layered strata in formations of limestone and chalk through-out France, Belgium, Germany, and England. In this region, Paleolithic Stone Age man displayed high levels of achievement in developing stone artifacts.

Flint is one of the many members of the silica family. Nearly all are exceedingly hard and durable. But flint stones possess a special quality in that they break into sharp-edged conchoidal flakes when struck or subjected to pressure. These shell-like flakes can then be readily adapted to form cutters, scrapers, arrowheads, hammers, and axes. Another rock, obsidian, also exhibits a conchoidal fracture, but it is found only in areas of volcanic activity.

In other areas, outside western Europe, these Old Stone Age people had to use other rocks, such as granite, diorite, andesite, and quartzite, if flint was unavailable. Although these rocks were hard, they did not exhibit conchoidal fracture and were, therefore, more difficult to fashion and less reliable in use.

During the second interglacial stage, there developed a new race called Pithecanthropus erectus, because they walked upright on their hind legs. They were rough, small-brained types who had discovered that sharp-edged stones were more useful in tearing apart animal flesh than were their hands and teeth. During later interglacial periods, other human cultures developed. Some produced very crude hand axes from flints. Others learned to remove the outer weathered crust from flint stones to improve the stock. Later it was learned that freshly excavated flints were easier to work than the tough-skinned nodules found among the stream gravels. This is where active mining for flints first began in earnest.

About 80,000 year ago, during the last interglacial period, Homo neanderthalensis appeared. These people lived in caves and wore warm clothing as the climate grew more severe. During their 60,000 years of existence, they further developed the use of stone implements to suit their improved lifestyle.

One of the earliest clues we have to the mining activities of the ancients is associated with their funerary habits. Primitive man believed in immortality. It was understood that blood was the essence of life, and to restore life after death it was necessary to provide adequate replacement of the bodily blood lost in death. This need was provided by the custom of burying the body in a mass of red ocher powder and with lumps of red stones scattered around the grave.

Red ocher is an oxide of iron called hematite by the Greeks: a word meaning blood stone. Hematite was therefore mined in substantial quantities for these funerary and other purposes. When surface deposits had become exhausted, this powdery ore was mined underground, as in Bomvu Ridge in Swaziland, southern Africa, before 40,000 B.C. (see figure 6.1). Bomvu Ridge is, therefore, the oldest known mine in the world. In 1957-58, the Swaziland Iron Ore Development Company was formed to explore this same deposit. Some 48 million tons of massive hematite ore, averaging 62 percent of iron, were developed. In August 1964, this mine was placed in production, some 50,000 years later.

It is not clear from present evidence as to the earliest time of original mining activity. Although flint stones had been used many millennia earlier, it can be supposed that these were obtained mainly by selection of stones (eoliths) lying on the surface or in the beds of streams. We do know, however, that in late Acheulian times the practice had developed of excavating flint from beds of limestone exposed in the banks of rivers or in hilly country. This may be assumed to be the earliest form of simple mining activity, about 50,000 years ago. Nevertheless, the mining of ocherous hematite iron ore in Bomvu Ridge also began up to 50,000 years ago.

Fig. 6.1. Map of Southern Africa.

Some time after the last ice age, about 35,000 years ago, the Cro-Magnons spread westward into Europe from Asia to seek better hunting grounds. They gradually replaced the Neanderthals and flourished in an environment of excellent flint deposits, with an abundance of caves and good hunting potential. They were an advanced race of people. One of their more useful developments was in flint-stone technology: they found that stones of flint produced much better and more uniform flakes or shells when pressure was applied than when struck. This was a significant advantage in that very thin, sharp flint blades could thereby be produced for tipping spears, darts, javelins, and fish hooks. Flint now became a luxury and represented the acme of mineral possession for thousands of years. Later on, these people developed fine tools and needles for stitching garments and for fishing. These were made from the bones of animals. The Cro-Magnons were a talented race; they developed some beautifully decorated art forms on bone and ivory sculptures with ochers and pigments.

The Old Stone Age came to a close about 8000 B.C., after covering about 300,000 years of man's early existence. During this time range, early man had invented most of the fundamental stone tools that were to serve his descendants for many thousands of years to come.

Men of the Old Stone Age had been the first to find and use the earth's primary mineral resources. At the close of the Paleolithic era, Cro-Magnon man was being assimilated by newer races invading western Europe from the Baltic region, from the eastern Mediterranean, and also from North Africa.

7 Neolithic (New) Stone Age

The Neolithic Stone Age was also called the Age of Civilization because this is when the earliest signs of organized society appeared, with several revolutionary changes in early man's lifestyle. Among the new arts developed were the polishing of his stone tools, the making of pottery, the domestication of wild animals, and the use of seeds and plants leading to the cultivation of land and the growing of crops for food. The nomadic way of life was thenceforth abandoned. A more settled existence brought about the establishment of villages, with a consequent increase of the population, which incidentally increased the demand for stone tools.

Flint had become an important and valuable natural resource. Those tribes possessing good workable deposits of flint experienced an industrial boom. Certain localities became famous for the quality of flint mined and of the implements produced. In England, Belgium, France, and Sweden, shafts were sunk and galleries driven into chalk and limestone deposits to mine the prized flint.

Crude mining tools consisted of rough picks and hammers of flint, and pointed stone with handles of stag horns. Picks of red deer antlers were used to loosen chunks of flint. Shovels were made from the shoulder blades of animals. Fire-setting was used to loosen up a working face of flint.

As Neolithic man turned to agriculture, his diet came to include more grains and vegetables and less meat protein. Salt became a necessity of life, both as a food intake and for religious ceremonies. Cakes of salt were later used as money (currency). Hence the term salary. Ochers and amber were employed for decorative purposes.

In this way, salt, amber, ochers, and good quality flints became increasingly important objects of trade. Most of the salt was mined in what is now known as Austria, at Hallstatt, near Salzburg (see figure 7.1). Amber was found on the shores of the Baltic Sea.

Meanwhile, the Middle East had become the known world's bread basket in these days. The extremely fertile land of Mesopotamia, in the river valleys of the Tigris and Euphrates, produced most of the food, in a warmer climate with longer growing seasons. This was probably the Biblical "Garden of Eden." Farming became highly advanced in this area.

It was natural that great trade routes should become established between western Europe and the Middle East and beyond during this period. These trade routes passed through Mesopotamia. Trade was effected with flint, salt, amber, bitumen, pottery, and agricultural products, and much later with ivory, spices, gold, and other metals.

Although France had the best deposits of flint, the greatest developments in civilized advancement during this Neolithic period occurred in the Middle East: in Mesopotamia, Persia (Iran), Media (Iraq), Arabia (Saudi Arabia, Syria, and Jordan); see figure 7.2.

Although flint and other stones had become the raw materials for fashioning tools, implements, and weapons, and hematitic ochers had been used in the powder form for decorative and funerary purposes, they were later to lead to the use of metals.

The first metal to attract man's attention (in about 6000 B.C.) was gold. Gold was first found glittering in streambeds or water holes as placer gold, and sometimes as free native gold in rock outcrops. Gold naturally collects in stream deposits because when pure it is 19 times heavier than water. But gold was found to be too soft for use as a tool or implement. It was valued for its great charm, its enduring, untarnishable yellow glitter, and its great malleability and ductility.

The purity, or fineness, of gold is expressed as the number of parts per thousand of pure gold in an alloy, or in terms of karats (not to be confused with carats, the unit of weight of gemstones). Pure gold is 1,000 fine or of 24-karat value.

When pure and uncontaminated with silver and other metals, it can be hammered into very thin sheets (gold leaf) in an art known to the ancients as "gold beating." Gold has been beaten to a thickness of 1/280,000 of an inch, although nowadays, gold leaf is only about 1/16,000 of an inch thick. Such gold leaf is used these days to cover the domes of public buildings, such as the state capitols in the United States, and the Grand Place in Brussels. It is highly regarded for this purpose, not only for its excellent decorative effect, but for its great resistance to atmospheric weathering.

As an internal decoration, countless works of art have been gilded with gold leaf, as in many of the cathedrals,

Fig. 7.1. Map of Central Europe.

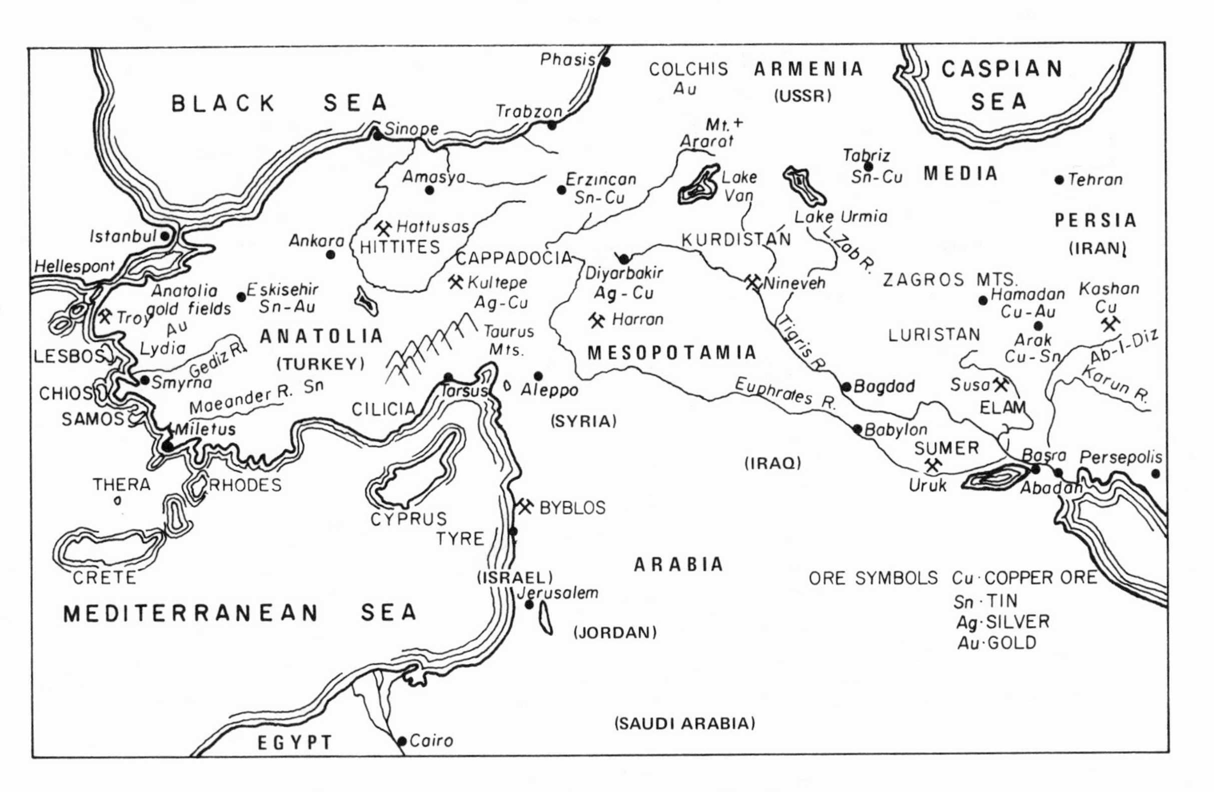

Fig. 7.2. Map of the Middle East.

museums, and palaces of Europe and South America. From earliest times, gold was highly prized by kings, and any gold found became the property of the kingdom.

One of the most fabulous collections of gold ornaments is now held in the famous Hermitage Museum in Leningrad. This particular collection represents some of the finest works of the goldsmith's art ever created. The little known race of marauding warriors, the Scythians, had apparently commissioned artisans from other lands to craft these objects. This work was done between 800 and 400 B.C. with gold won from the Altai Mountains in Siberia.

Gold is also one of the most ductile of metals. A single ounce of pure gold can be drawn into a wire 50 miles long before breaking. Such wire is used in making gold lace. It is also woven with silk threads into cloth (known as "cloth of gold"), highly prized for ceremonial trappings and costumes.

And so the ancients were highly attracted to this new-found metal (gold), not because of its usefulness for making tools and weapons, for which it was too soft, but because of its very great ornamental value and for personal adornment. It was, therefore, of considerable value as an object of trade, either in the raw form or fabricated.

As a medium of trade, gold later developed utility in the form of coinage, to facilitate the practice of bartering. It is still regarded as the ultimate store of real value.

In many of the royal burial grounds of the ancients, fabulous treasures in the shape of gold and silver ornaments, richly decorated with gems, have been unearthed. Such discoveries are (a) the Royal Cemetery of Ur (Sumeria), dating from 2500 B.C., (b) King Tutankhamen's tomb in Egypt, 1350 B.C., and (c) tombs of other Egyptian pharaohs.

The knowledge of gold mining and the ancient art of goldsmithing travelled from Nubia and the Nile Valley to many cultures through the Middle East and Far East in those days; it also spread to western Europe and the Iberian Peninsula. The gold deposits in Nubia were fabulous. Egypt became one of the richest countries in the known world, and made great contributions to early civilization for a period of over 3,000 years, from about 3300 B.C. From about 3300 B.C., Egyptians were recovering turquoise and other decorative minerals from deposits in the Sinai Peninsula.

In about 1500 B.C., upon the death of Thutmose I, his daughter became the first woman ruler of Egypt. Queen Hatshepsut (1526-1482 B.C.) sought new sources of wealth to decorate a temple for her father and herself. She set up a new expedition by sea in about the year 1490 B.C., to the land of Punt, where fabulous resources of gold, ivory, ebony, rare animal skins, and stibnite were known to exist. Similar expeditions had been mounted previously in about 2700 and 1900 B.C. to this land of great riches, whose geographical location can still not be pinpointed; it is now assumed to be either in southern Arabia, in Somalia, or in Mozambique. The queen's successful adventure is represented by inscriptions and colored reliefs of her magnificent temple near Thebes.

In the Nile Valley, a place called Coptos was the world's first gold boom town. It was in the general area of Wadi Hammamet. Placer gold had been washed down the water-courses from the gold-bearing veins found later in the granite hills above. The ruins of over 1,100 stone huts of the miners may be seen at Foakhir.

The world's oldest mine map, made from papyrus, is held in the Turin Museum (Museo Egizio di Torino). It shows the huts of these Egyptian miners, the road to the gold mines, and the hills within which the gold veins occurred.

The gold-mining areas of Egypt occurred in a strip east of the Nile and west of the Red Sea extending southward from Coptos for about 500 miles into the Nubian desert (see figure 7.3). Both placer and vein gold were mined, the richest deposits being worked in Nubia. In the old Pharaonic language, the word nb meant gold, and nubia, a goldfield.

By about 1300 B.C., underground mining of vein gold was well-established in Nubia, under Egyptian control. There were over a hundred mines in the area. The earliest mines were merely trenches and pits, but later galleries and adits were driven into the hillside, and then inclined shafts were sunk. The deepest mine was 292 feet deep and extended for about 1,500 feet along the vein. The roof was supported by pillars or by wooden props. The broken ore was removed from the mine in baskets. The principal tool was a stone hammer. A chain of forts was required to protect the flow of Nubian gold along the rich trade routes.

Egypt became the dominant power in the Middle East, mainly because she had the greatest gold-filled treasury in the ancient world. Thebes became the capital, the most magnificent city in the world during Egypt's greatest period of power and achievement.

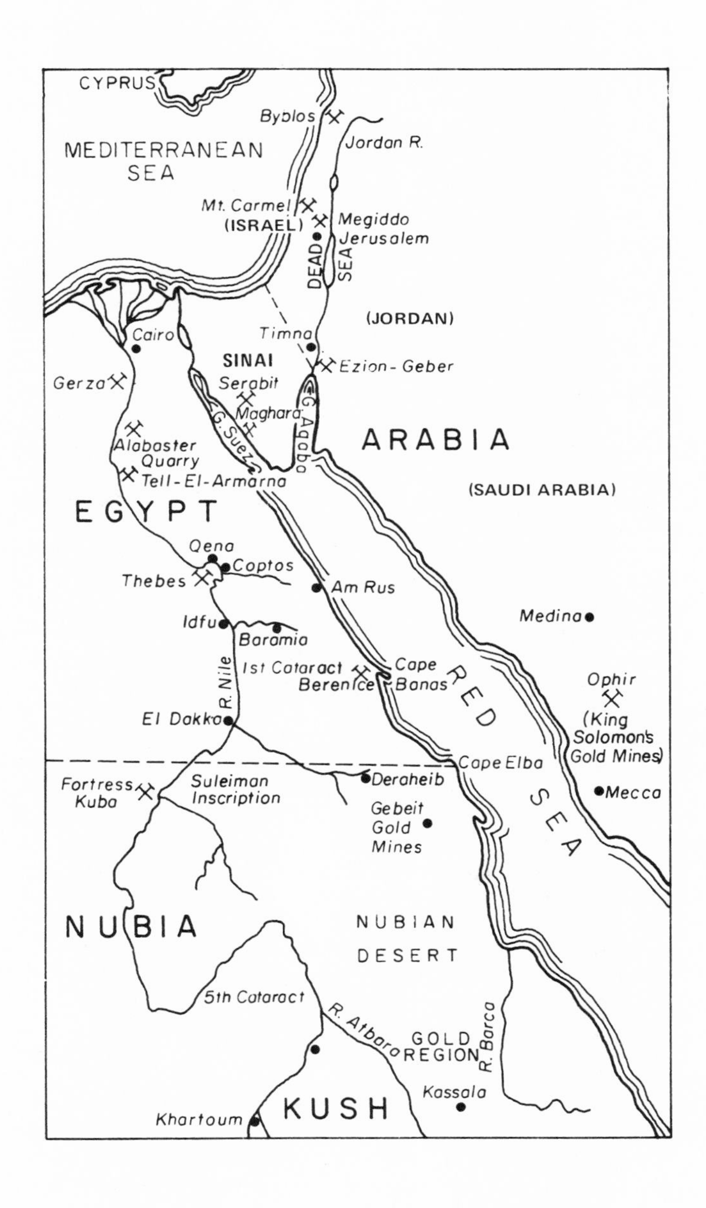

Fig. 7.3. Map of Egypt.

8 The Copper Age

The Copper Age commenced in 5500 B.C. This was the first metal to be named for one of the Ages of Man.

Oxidized copper minerals such as the carbonates (the green malachite and the blue azurite) had been found earlier, before the days of smelting, but the ancients did not realize that these minerals contained metal, and therefore they had no particular regard for them for fashioning their tools and weapons. These brightly colored minerals, along with turquoise, talc, and ocher had been used purely for ornamental purposes, including the glazing of pottery.

It was only after native copper had been found in these deposits that the ancients realized this was a metal that was harder than gold, and therefore had possibilities for use as tools. However, although more readily fashioned than flint, it was also much less durable and soon lost its "edge" in use.

In the Middle East, Neolithic man, while seeking suitable stones and also collecting gold, came upon occurrences of this reddish metal, which was really native copper. Although relatively soft, it could be hardened by hammering. But after a time it was found that increased hammering made copper brittle. This deterred the ancients, until by accident it was discovered in 5000 B.C. that a lump of hardened copper dropped into a fire became softer and less brittle, more malleable and easier to shape. This process is now known as annealing.

They later found that copper could be liquefied in a strong fire. This inspired the ancients to melt copper and pour it into shaped molds in the form of castings. It was thereby solidified into tools of the desired shape.

Following this development, attempts were made to turn the green and blue carbonates of copper (malachite and azurite) into copper metal by heating them in a strong wood fire which developed its own charcoal. This was the very early beginning of copper metallurgy, the smelting of copper ore, in about 4300 B.C. It developed in many parts of the Middle East, including Iran, Turkey, and the Sinai Peninsula (see figure 7.2). Actually, a similar evolution of copper working and smelting developed in North America after 4000 B.C. by primitive Indians in the Great Lakes region. This old copper culture was probably introduced from Siberia through Alaska.

The probable birthplace of copper smelting is at Susa, Iran, where the art of pottery making and that of annealing and casting was already highly advanced for those times. Later, these and other peoples from Armenia settled in the fertile valley of the river Tigris, at the head of the Persian Gulf in a territory known as Sumer. This land included an area that was probably the Garden of Eden. It is now arid and desolate, and mostly covered by sand dunes. It was much later to be known as Babylonia.

Following the Great Flood of 4000 B.C. (according to the Gilgamesh Epic), the delta of the Tigris became silted up, and new cities were established by a race known as the Sumerians. They were an advanced people and were skilled in many arts including metalsmithing.

Most of the gold used by the Sumerians came from Iran, or was traded from northern Armenia between the Black and Caspian seas. This later included the land of Colchis, which was the legendary home of the Golden Fleece, sought in about 2500 B.C. by Jason and his Argonauts. This legend derived from the use of woolly sheepskins to collect fine gold particles washed down the streams.

The Sumerians are recognized as the early developers of civilization. With their creative talents, they developed agriculture, irrigation, metallurgy, commerce, mathematics, and religion in one of the first empires of the world. Their history has been passed down to us in pictograms written on clay tablets. This civilization probably reached its zenith in 2100 B.C., known as the Golden Age of Ur.

At about this time, the Egyptian civilization was also well under way. Metalworking had been brought to the Nile in 3500 B.C. by skilled immigrants from Sumer. These were known as the Gerzeans, because they settled in Gerza on the Nile. Over a long period they were assimilated by the Egyptians, and their culture was similarly absorbed.

The tombs and monuments of the early rulers of Egypt had been built from mud bricks, in a similar style to that of Mesopotamia, during this period. The excellent stone of the Nile Valley was not used in architectural construction until the use of copper chisels was developed for cutting and shaping these huge stone blocks.

In about 2900 B.C., Egypt began to use her large deposits of massive stone to build monuments such as tombs and pyramids, in honor of her godlike kings, during the Third Dynasty. Egypt possessed an abundance of durable limestone, granite, diorite, and alabaster.

By now, copper had become widely available and had been introduced in the form of stonemason's tools. These made possible the accurate cutting and shaping of stone blocks for pyramid construction. The first pyramid was built by King Djoser in about 2800 B.C. at Saqqara, near Cairo. It still stands today, to a height of 190 feet above the desert. It is the oldest standing structure in the world.

But the largest is the Great Pyramid at Gizeh. This one is approximately 756 feet square at the base and covers 13 acres. About 2,300,000 stone blocks averaging 2.5 tons in weight were quarried to construct it to a height of 470 feet. Some blocks weighed as much as 45 tons. It is regarded as one of the Seven Wonders of the Ancient World.

Apart from the pyramids, other great structures and edifices of stone were erected all along the Nile, not only as massive blocks, but also in the form of artistically sculptured statues. Egyptians had become the stonemasons of the world. Several Egyptian obelisks now stand in other cities, such as Rome, New York, Paris, and London. Egyptian carving in alabaster is called "onyx marble." Translucent peach-colored alabaster has been carved into jars, vases, and figurines of outstanding beauty. Exquisite articles carved by Egyptian craftsmen are now admired in many of the world's famous museums.

The Egyptians derived their copper ores (mainly malachite and azurite, along with turquoise, a complex phosphate of copper and aluminum) from Wadi Maghara in the Sinai Peninsula, east of the Gulf of Suez. An expedition was led to this deposit in about 2500 B.C. following invocations to Hathor, the Egyptian goddess of joy and good fortune, the Lady of Turquoise, and the moon goddess of silver. There was a widespread belief that silver was developed by the moon's rays (the magic of the silvery moon), just as the sun's golden rays were believed to be the active agent in the development of gold.

Later, an important mineral deposit was found at Serabit el Khadim, in Wadi Maghara. Here an extensive shrine about 250 feet long was built in honor of Hathor. Rock inscriptions and carvings in this area show that these mines were in use by Egypt for over 2,000 years.

These deposits were not really large, but the demand for copper was correspondingly not particularly great. Copper was used mainly for ornamental purposes and for making tools for cutting and shaping of stones at that time.

The copper ores were smelted in the Sinai district and large heaps of smelter slag may still be seen there. Charcoal was used as a fuel. For this purpose, shrubs and trees had to be cut from a wide area. The smelting process was crude, as evidenced by the discarded slag, which contained 2.75 percent of copper.

In about 1167 B.C., after Ramses III died, Egypt became disorganized, and foreign invaders came in turn from Ethiopia, Persia, and Greece, followed by Romans, Arabs, and Turks. The last record of copper mining by the Egyptians in the Sinai Peninsula was about 1150 B.C. After this time, most of the copper used in Egypt came from Cyprus. Although the Bronze Age had long since begun in the Middle East, Egypt, probably because of the scarcity of indigenous tin supplies, had been slow to embrace the use of bronze.

But Egypt had given great gifts to mankind in stonemasonry, architecture, agriculture, geometry, medicine, jewelry, metalsmithing, sculpturing, glassmaking, papermaking, and fine weaving. All these arts were handed down to civilization through the Phoenicians, Syrians, Hebrews, Cretans, Greeks, and Romans. In many ways, the Egyptian civilization ranks as the greatest known to mankind. Gold had been a significant force in building up her civilization; and the use of copper tools for trimming blocks of stone made possible the building of her great pyramids and monuments.

9 The Bronze Age

The Bronze Age commenced in about 3000 B.C., perhaps in Luristan, now part of Iran. No event in the history of ancient man had such an impact upon his cultural development as the discovery of a mineral known as cassiterite (tinstone). For ages, relatively pure copper had been the sole working metal of mankind. But it was soft and had to be hardened and sharpened by constant hammering. Yet if hammered too much, it became brittle.

By adding a small percentage of tin to their copper, the ancients produced a metallic alloy much harder than copper. It gave a more fluid melt and was easier to cast. This became known as bronze.

For over 2,000 years following its discovery, bronze was the cornerstone of the world's industry and art. Various types of bronze made from different percentages of tin in the alloy were used to fashion a wide assortment of tools, weapons, and other products.

Tin occurred chiefly as the oxide mineral cassiterite. But comparatively few deposits were found in the early days of the Bronze Age. These have long been worked out. Perhaps some of the early supplies were obtained from the Far East: from Thailand and Malaysia. Tin became a regular article of trade between countries. In about 2000 B.C., early Iberian prospectors, probably forefathers of the Basques, arrived in Wales seeking tin supplies. Some crossed into Ireland and discovered large deposits of alluvial gold and much copper, but no tin, which by then had become a strategic metal for bronze production. Later the Irish tribes found tin in Cornwall.

It is known that the Phoenicians were based on Tyre (around 1200 B.C.). They established colonies at Cadiz (about 1100 B.C.), in Carthage (814 B.C.), and in Cartagena,

Spain (225 B.C.) (see figure 9.1). They secured supplies of tin from Spain, Portugal, and Cornwall, after 600 B.C. Later, the Greeks and Romans did likewise. The need for a steady supply of tin and lead ores was the main reason for the Roman invasion of Britain in 55 B.C. and its occupation in 43 A.D.

Bronze was a much tougher metal than copper. Through its use, daggers and swords were fashioned, thereby advancing the science of warfare between tribes. Beautiful art forms in bronze were also accomplished.

But bronzemaking reached western Europe a thousand years or so after its development in western Asia. It played an important part in the development of industry, commerce, and art and brought western man from the darkness of the Stone Age into the Age of Metals. Metals such as copper, tin, lead, and gold now came into expanded use, and inspired man to attempt new discoveries.

Following the destruction of Babylon in 1758 B.C. by the Hittites, the Assyrians rose to power and conducted sound trade arrangements with the various races in Anatolia (Turkey) to gain supplies of metals in exchange for their woven fabrics of linen and wool (see figure 7.2).

Various regions of Anatolia had rich deposits of lead and silver, and are credited with the earliest practice of lead smelting. Most of the silver was obtained by crude cupellation of argentiferous galena ores. Silver discs (shekels) came to be used first as a unit of weight and later as coinage. Metallic lead had a limited use for many centuries because of its extreme softness.

Anatolia produced significant mineral wealth, thereby stimulating the development of art, industry, and commerce, because it was astride the main trade routes of the camel and donkey caravans. The Babylonians and Assyrians were able to stimulate their own economies by importing supplies of metals from Anatolia as raw materials for their craft industries.

Following the beginnings of the Bronze Age in western Asia, its influence spread slowly to Europe. There were two main routes. One extended up the valley of the Danube from the Black Sea. The other made use of the island of Crete as a stepping-stone from both Egypt and Asia Minor to Europe via Greece.

Crete had for several millennia developed a civilization of its own (the first in Europe), ranging from the Stone through the Copper and the Bronze ages, based upon transfusions of craftmanship from both Egypt and Anatolia. But Crete had very few indigenous minerals. It was not the possession of mineral wealth that made Crete a great nation. It was her strategic trading position along the sea routes and her expertise in metalsmithing that led to her prosperity, coupled with the fact that, as an island nation, she had no defense problems. Her famous palaces, as at Knossos, Mallia, and Phaistos were luxurious treasure houses, but not fortresses.

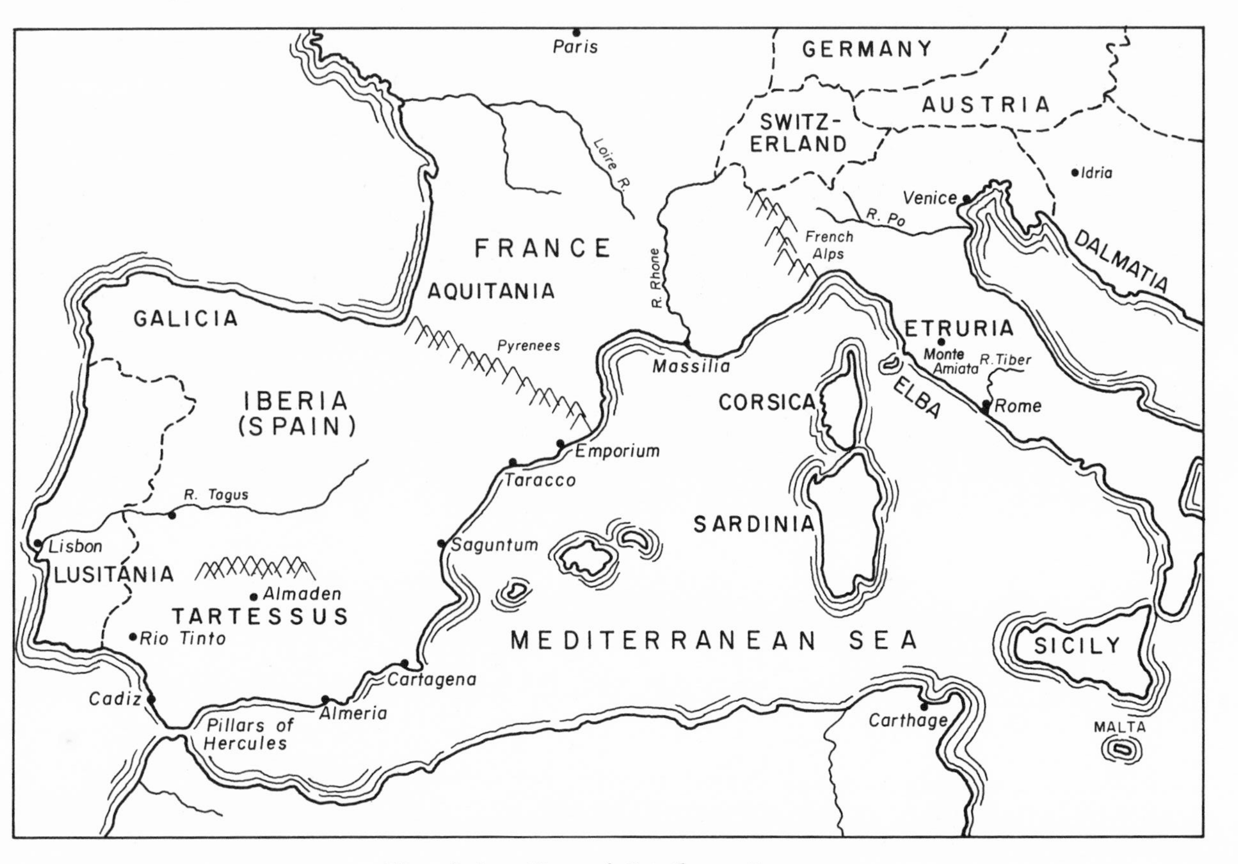

Fig. 9.1. Map of Southern Europe.

The kings of Crete built large merchant fleets which had command over the Mediterranean and made this island the earliest maritime power in history. Their ships linked the civilizations of the Middle East to the barbarians of Europe.

The essence of the power and culture of Crete was its long-standing trade and prosperity, based upon the originality of its art and the development of its expertise in metal-smithing, especially during the height of the Bronze Age. In about 1450 B.C., however, Crete was stricken with some sudden catastrophe, perhaps a violent earthquake or volcano. Marauding looters from abroad then took advantage of her defenseless position and her cities were totally destroyed. But Crete's culture now became transferred to Greece, where a new city, Mycenae, had arisen. This was the birthplace of Greek civilization, in about 1700 B.C. (see figure 9.2). Its inhabitants had earlier filtered southward from the Balkans and the Caucasus Mountains through Thrace and Thessaly. Then about 1300 B.C. another wave of migrants crossed the Dardanelles from Troy and entered Greece, bringing Anatolian culture with them. These were known as Greeks. They assimilated the inhabitants and developed the economy. The Mycenaeans built merchant fleets and traded successfully around the eastern Mediterranean, and especially with Cyprus.

The island of Cyprus possessed an abundance of copper. It played a large part in the emergence of man from the Stone Age. Copper metallurgy came to Cyprus in 2600 B.C. but it had been the demand from Crete that developed her copper resources after 1600 B.C. Copper ingots were also supplied to the industries of Anatolia, Syria, and Egypt. Later, the copper mines of Cyprus were operated by the Mycenaeans. Today they are again being worked, by the Cyprus Mines Corporation.

The Mycenaeans had developed artistry in bronze to the peak of perfection, not only in ornaments, but also in weapons such as swords with flanged hilts. These added to their military supremacy and assisted in their conquering of Troy. But in about 1100 B.C., the Mycenaeans were subjugated and absorbed by new invaders.

Meanwhile, the bronze culture had been diffusing into Europe through Italy and Spain. Although Spain possessed vast resources of copper, tin, gold, silver, and lead, she had not learned about metallurgy. Therefore, the early Spaniards had to rely upon finished articles traded to them by the Aegean peoples from 2000 B.C. Later, they were to mine and smelt their own ores and fabricate their metals.

The early Bronze Age also reached central Europe at about this time via the river Danube. This encouraged Europeans to search for ores locally and to smelt them by the use of metallurgical techniques developed in the Middle East. As a result, mineral deposits were found and developed in

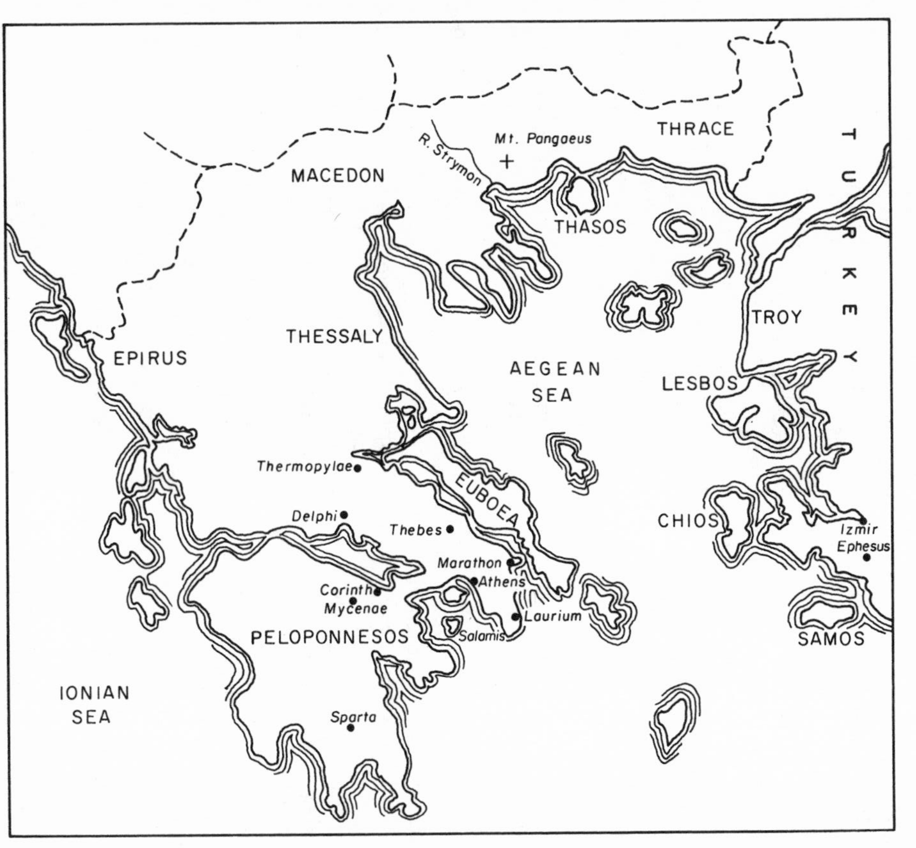

Fig. 9.2. Map of Ancient Greece.

Austria and in Bohemia. These areas provided copper and some tin, silver, lead, and gold.

About 1600 B.C., in the Unetice region west of Prague, Czechoslovakia, a very proficient culture developed in bronzesmithing. With small deposits of copper and tin discovered in Bohemia and with rich deposits to the south, these people perfected casting of bronze tools, swords, battle-axes, and helmets, and made fine golden jewelry. All these articles were available for trade because this region lay in the path of the "amber trade route" from the Baltic to the Adriatic.

Bronzemaking throughout Europe became highly intensive in the wake of the Bell-beater Folk who traded in roaming bands all over the continent. They set up regions of industrial activity at the junctions of recognized trade routes, with highly skilled metalsmiths, craftsmen, and potters as members of their teams. The Beater Folk blazed the trails of commerce all over Europe throughout this period, and into the British Isles in about 1500 B.C. This tribe originated in Spain.

Copper was mined in Cheshire and in Anglesey about this time and lead was mined in the Mendip Hills and in Derbyshire, England. But it was the Ueticians who later dominated southwestern England and came into possession of the rich copper and tin deposits of Cornwall. They imported expert craftsmen from Bohemia and developed their own style of bronze ornaments and jewelry. They thereby developed trade and commerce throughout the length and breadth of Europe.

And so the Bronze Age played a prominent part in the evolution of mankind from 3000 to 1400 B.C. It developed in the Middle East many centuries before its arrival in central Europe; but it eventually brought Europe out of the Stone Age, and taught Europeans how to make good use of their natural resources in the shape of minerals other than flint.

Not only copper and tin, but also the use of silver, gold, and lead expanded during this period. A process known as cupellation was developed in the years between 3000 and 2500 B.C. This was used to separate silver from lead. With a blast of air the lead was oxidized to litharge, and absorbed by the bone ash material of the cupel vessel; the silver remained. Lead metal was later recovered. This method was later developed to separate silver from copper by adding lead metal which preferentially alloys with the silver.

Mankind received a great impetus through the intellectual development of the arts and sciences, and in fact, all manner of cultural achievement.

10 The Iron Age

The many industrial uses of bronze (for manufacture of tools, implements, and weapons) were to give way to the development of a new metal, iron, in about 1400 B.C. Iron (or steel) is now the cornerstone of man's industrial existence. It covers a great range of his material needs.

After aluminum (8 percent), iron is the most abundant metal, in combined form, in the earth's crust, representing 5 percent by weight. It is spread very sparsely throughout the rocks and only when it is found in deposits sufficiently concentrated by nature is it rich enough to warrant extraction by mining. But iron ore deposits, mostly as oxides, occur widely across the world.

Although iron ores (ochers) had previously been mined as far back as perhaps 50,000 B.C. at Bomvu Ridge, these oxide materials had been used only as powders and pigments: for eye shadow, lipstick, and as war paint; for mural decorations; and for funerary purposes. Metallic iron was not known.

The earliest form of metallic iron used by man was of celestial origin, in that it was recovered from meteorites lodging in the earth's surface. Meteorites generally contain over 7 percent nickel and some cobalt. Otherwise, it is remarkably pure and malleable. It could be shaped to advantage into swords and daggers, and was first highly prized for this purpose. There was a limited supply, however.

At a later stage, the metalsmiths began experimenting with iron of terrestrial origin: naturally occurring iron oxides mostly. This work was done mainly by the Chalybdes, a tribe on the southeastern coast of the Black Sea. Following the practice of copper metallurgy, they smelted pieces of iron ore in a strong wood fire. Sufficient charcoal was necessary to surround the iron ore and to shield it from the atmosphere.

The temperatures attained were not high enough, however, to produce molten iron. Instead, a loosely coherent pasty sponge iron was produced with much slag in the pores. By hammering this mass while still hot, the slag was separated and a lump of reasonably pure iron metal was obtained. The remaining metal was hammered further to improve its quality and to produce ingots of iron or useful implements. Because of the hammering required, it became known as wrought iron.

Normally, this early so-called smelting process produced soft malleable wrought iron which could not be hardened because it contained too little carbon. But if kept in contact with charcoal for a longer period, the iron was found to absorb carbon (up to 1 percent). This carburized iron could then be hardened by raising it to a high temperature and cooling it rapidly by a technique known as quenching. If it became so hard that it developed a brittle structure, it could be tempered by reheating to a low temperature, thereby making it less brittle.

The judgment and skill of early blacksmiths was very important in producing the desired grade of iron. They gradually developed the art of making sword blades with this carburized tempered wrought iron.

The early techniques for making wrought iron implements by the blacksmiths of the day were rather weird. One school of sword makers believed that the best sword blades had to be quenched by plunging them when redhot into the body of a Nubian slave. Later, a pig was used. Other early authorities claimed that radish juice mixed with the juice of earthworms was required for the quench.

The smiths had mystical or magical qualities in the eyes of the common people, because of their skill with the working of metals. They enjoyed a social status just below that of the priesthood.

It was the Hittites who first began to use iron to make weapons, thereby introducing the Iron Age. The Hittites were originally nomads. They arrived in Anatolia about 2000 B.C. and within a hundred years had begun to overrun eastern Turkey. They were great horse breeders. Their economy was based on agriculture, but they also mined some copper, silver, lead, and iron from the mountains. Probably this was the first iron to be smelted (about 1400 B.C.).

About 1385 B.C. the Second Hittite Empire rose to power, largely because of the superior weapons of iron they had fashioned. The Hittite Empire became the foremost military and political power in western Asia. But the secret of ironmaking soon leaked out to other parts of the known world. This led to the disintegration of the Hittite Empire following the competitive challenge of neighboring peoples who had now acquired the art of ironsmithing.

The Assyrians, who had invented the chariot, defeated the Hittites in about 1300 B.C. They also captured Babylon in about 1100 and again in 689 B.C.

About 1200 B.C., the Philistines came in from the sea with their new iron implements of warfare and soon conquered the Israelites and developed the state that became known as Palestine. They specialized as blacksmiths, using imported iron. And so the Iron Age brought many economic advantages to Palestine. Later, the Israelites subdued the Philistines and King David reigned from 1013 to 973 B.C. in Jerusalem. His son Solomon developed an alliance with the Phoenicians.

Iron smelting was firmly established in Greece by 1000 B.C. Iron mines were operated by the Etruscans in about 900 B.C. in both Tuscany and on the island of Elba. Later, iron metallurgy was introduced to central and western Europe, between 900 and 400 B.C. This was known as the Hallstatt era, during which Iron Age artifacts were fabricated from the smelting of local iron ores and first identified as such.

These famous iron mines were at Noreia, about 40 miles from Hallstatt. They produced "Noric iron" from which the celebrated Noric swords were produced. This iron did not need tempering, and these swords were greatly prized by the Romans.

Hallstatt was, therefore, an important center during the early centuries of the Iron Age. It was also the site of prehistoric salt mines worked first by Neolithic tribesmen, followed by Bronze Age Romans and medieval Christians. These old salt mines are in the mountains of Salzburg, Austria, many thousands of feet above the lake, near Hallein and Gmunden, the gateway to the great salt route established in medieval times. Many operators here became rich, not as salt miners, but by collecting salt taxes or fees for shipping salt to the big barges plying down the Traunsee to the Danube. A famous prehistoric cemetery has been found in this area.

In the 6th century B.C., iron metallurgy had spread to Spain, where the Catalan furnace was developed; this used two pairs of bellows for the draught. Between 500 and 450 B.C., successive waves of Celts invaded Britain from the European continent in search of mineral wealth. Using advanced weapons of iron, they had no difficulty in subduing the English tribes and in dominating the tin trade in Cornwall.

KING SOLOMON'S MINES

The Phoenicians had migrated from the Persian Gulf or Babylonia as early as 2750 B.C. and established thriving seaports along the coasts of Israel and Lebanon. They were expert metalsmiths and traders. Tyre especially became an opulent Phoenician city-state.

King David of Israel, and his son Solomon, entered into friendly trade relations with powerful King Hiram of Tyre, in about 950 B.C. There were few artisans among the Israelites, so Solomon had to engage Phoenician craftsmen and materials to build his temple in Jerusalem.

Solomon's temple was supported by carved cedarwood columns overlaid with gold. His palace contained great halls lined with statuary and its walls and ceilings were of cedar decorated with gold leaf. No only really knows the source of the gold and other materials used.

But on the western side of Wadi Araba, about 15 miles north of Eilat, at a place called Timna, copper is still being mined today. This was the world's first large copper belt. Tall, red sandstone cliffs known as the "Pillars of Solomon" guard the entrance to "Solomon's Copper Mines" in the Valley of the Blacksmiths. Apparently, over 80,000 slave miners here worked small surface exposures of rich copper ore, just as the Egyptians had done in the Sinai Peninsula to the south over 300 years earlier.

Some of King Solomon's copper smelters were located in Jordan, east of Timna, at a place called Ezion-geber (see figure 7.3). They had apparently been operated earlier by the Phoenicians. Recent exploration, however, indicates that copper smelting had been carried out at Timna in three major periods: first, in the Chalcolithic era, around 4000 B.C.; second, between 1400 and 1200 B.C., by the Egyptians and perhaps by the Phoenicians; and third, by the Romans in the 2nd century A.D.

Much of the copper produced in the Ezion-geber smelters was bartered for gold, ivory, perfumes, spices, and incense from Asia, traded along the great inland trade routes. A maritime trade route was also set up by King Solomon and King Hiram of Tyre, who built their own navies, based on Ezion-geber, an important outlet to the Red Sea and the Indian Ocean.

To this port, large shipments of gold were discharged for King Solomon. Some believe the gold came from Zimbabwe. But it probably came from the land of Ophir, a land of prodigious legendary wealth. Where was Ophir? This is a great mystery. It was not the land of Punt, because the Ophir gold was particularly pure, whereas the gold from Punt had a slight greenish tinge due to impurities of copper or antimony. Some believe that Ophir was in Yemen in the region of Saba, ruled by the Queen of Sheba. Saba was extremely rich in gold and silver. More recent studies show that Ophir was further north in Saudi Arabia, at Mahd adh Dhahab (meaning "cradle of gold" in Arabic), about midway between Mecca and Medina.

At any rate, the ships of Hiram and Solomon, returning from India laden with spices, sandalwood, ivory, peacocks, and gemstones, picked up incense from Aden, and finally gold and silver from some Arabian port. The Ophir mines are reputed to have produced 31 metric tons of gold, about one-half of the known gold supply of the ancient world. About one million tons of discarded waste rock have been found in the Mecca-Medina area (see figure 7.3). In 1979, the author visited this ancient mining area, now about to be resurrected by Consolidated Ltd.

But Solomon's ivory throne, the gold from Ophir, and the wealthy economy of Tyre, whose streets were said to be paved with gold, attracted foreign invaders. The Assyrians had long sought an outlet to the Mediterranean Sea. Phoenicia and Israel had stood in the way. Several waves of Assyrians and Chaldean invaders came between 853 and 586 B.C. when they were finally victorious. King Nebuchadnezzar of the Chaldeans had defeated the Assyrians and then, in 586 B.C., devastated Jerusalem and the Temple of Solomon. Ten thousand Hebrews and thousands of Phoenician craftsmen were led in chains to rebuild and embellish great palaces and public buildings in Babylon. The erection of the famous bronze Ishtar Gate, the Tower of Babel, and the Temple of Marduk, surmounted by a shining pinnacle of gold, were thus built by slaves. Nebuchadnezzar's Palace and the Hanging Gardens became famous, richly decorated with ivory, gold, and silver. But in 539 B.C. Babylon was taken by King Cyrus and became a province of Persia.

THE PHOENICIANS IN SPAIN

Meanwhile, the great body of Phoenicians escaped by sea from Tyre to Carthage where they set up a new base in 814 B.C. in what is now known as Tunisia. From here they had commanding power in the Mediterranean; they had earlier established colonies in Malta, Sicily, Sardinia, and at Cadiz and Malaga in Spain to secure mineral supplies for their craftsmen back home. By 300 B.C., Carthage was the richest city on earth. Most of her wealth had been derived from the mineral deposits in the south of Spain.

West of the Spanish Guadalquivir River to the Portuguese border and south to the coast west of Cadiz lies the great pyrite belt from which copper, silver, and gold have been mined for nearly 3,000 years.

The oxidized copper minerals of malachite and azurite, derived from this cupriferous pyrite, were overlain by rich silver ores which were previously held by the Tartessians. The Phoenicians from Tyre had settled in Cadiz in about 1000 B.C. and monopolized these silver deposits to develop their commercial trade around the Mediterranean and to supply their industries at home. There is some evidence that King Solomon of Israel shared in the exploitation of these deposits.

These early operators smelted copper and silver ores and separated these metals by adding lead ore to the furnace feed. This was known as the liquation process. As the lead formed an ingot at relatively low temperatures, it carried the silver with it. These metals were later separated by the cupellation process. On heating, the lead oxide (litharge) was absorbed, leaving silver behind. Lead was found in many small veins in southern Spain but extensive deposits were developed by the Carthaginians in the 3rd century B.C. at Cartagena. (In 1976, the author visited lead mines still working at La Union, six miles east of Cartagena.)

Diodorus Siculus wrote that up to 580 B.C., Phoenician ships sailed from the city of Tartessus, heavily laden with bullion, and that even their iron anchors were replaced with those made from silver. But during the siege of Tyre by Nebuchadnezzar, there was a blank period of activity by Phoenicians in southern Spain. Meanwhile, the Phocaeans of Greece took advantage of this hiatus. They bartered wine and olives for metals in Tartessus until this city was destroyed by Carthaginians in 510 B.C.

THE GREEK CIVILIZATION

The first Grecian tribes arrived from northern areas in about 2000 B.C. These various tribes later came to be known as Achaeans. They were expert metalworkers in gold, silver, copper, and bronze and made Mycenae the most powerful Greek city by 1550 B.C. But the Achaeans were subdued in about 1100 B.C. by the Dorians whose weapons of iron were superior.

As a result, great changes in the development of Greece were to ensue. The Dorians had captured all of Greece except Attica, the area near Athens (see figure 9.2). They made their capital at Sparta. There were plentiful supplies of iron ore in Greece and industry expanded accordingly. Greece later developed her literature, philosophy, science, art, and architecture.

Meanwhile, following the Dorian invasion, some of the Achaeans (Myceneans) fled to Asia Minor and established the Ionian League, based upon the city of Ephesus. In this region, large deposits of metallic ores of silver, tin, and gold occurred, in a strategic position astride an important trade route. Ephesus became a flourishing commercial city. The Ionians became renowned for their craftsmanship in silver. Ephesus was the birthplace of Ionic forms of architecture.

About 60 miles to the north, the ancient city of Smyrna (now Izmir) was built in 1015 B.C. It was later absorbed by the kingdom of Lydia. Here, coinage was invented to replace the clumsy system of bartering. In 680 B.C., coins were stamped upon discs of electrum (a natural alloy of gold and silver). In 560 B.C., coins of pure gold and pure silver were used to stimulate trade and to develop banking.

The idea of coinage then spread to the mainland of Greece, where the system of currency was rapidly developed in the 5th century B.C. This was made possible by very rich gold and silver deposits in the Thasos and Siphnos islands and at Damastium (in Epirus), founded in 627 B.C.; and also by the mining of extensive deposits of silver-lead ore at Laurium.

These Laurium deposits, situated about 25 miles south-east of Athens, were first worked in the 15th century B.C. and reached their peak about a thousand years later. The ore of Laurium contained an argentiferous galena which was found impregnating the limestone rock in a horizontal tabular deposit up to 35 feet in thickness, not over 350 feet below the hilly surface. The mines were owned by the city-state of Athens but operated under lease by private citizens who paid a royalty of about 4 percent of the value of production.

Both open cut and underground mining were undertaken but mostly the latter. Excavations have revealed extensive underground workings. More than 2,000 shafts had been sunk into the ore body, to a maximum depth of 386 feet, but always above sea level. The main shafts were about 4 by 6 feet in section, with footholds in the walls for climbing. Some shafts were inclined to act as stairways. Underground drifts were restricted in size; the miner had to work in a crouched position. Hammers, picks, chisels, and shovels were made of iron with wooden handles. Very little timber was available for roof support, and pillars of ore were used extensively. The workings were illuminated by oil lamps. Ore was hoisted to the surface by a windlass or carried in leather baskets up inclined shafts.

At the surface, the waste rock was hand sorted from the ore. With iron pestles in stone mortars, the ore was crushed by hand, or in stone rolling mills operated by slaves. When crushed, the ore was spread upon sloping tables called <u>laveries</u>. Water gently flowing across the laveries washed the <u>gangue</u> materials away as tailings, leaving a concentrate of silver-bearing galena. The concentrates were then dried and roasted in small, round cupola furnaces to remove the sulphur. The roasted concentrates were reduced to metallic lead together with the silver. The silver was separated from the lead by cupellation. The lead became oxidized to litharge which was collected for later recovery of the metal by smelting with charcoal. The silver was melted and cast into bars to be sent to market or to a mint for coinage. Lead was used mainly as a mortar for grouting iron rods or dowels in the jointing of stonework. The production from these mines formed the economic support of the city-state of Athens for many centuries. Since 1860, the mines have been reopened by a French company.

Another city established by the Ionians in about 1000 B.C. was Miletus, in an area rich in mineral resources. By the 6th century B.C., Miletus became the richest city in the Greek world. Wealthy merchants sponsored art and borrowed from Babylon and Egypt for cultural inspiration.

The Greeks also expanded westward along the Mediterranean. They formed colonies at Nice and Monaco. In 750 B.C., they settled in Sicily and near Naples. Here, their art and philosophy spread throughout the region (see figure 9.1). About 600 B.C., they founded Massilia, which became an important port to the trade route up the Rhone River. This is where Marseilles now stands. This route continued northwestward along the river Loire to its mouth, giving an important connection to Cornwall, England; and thereby circumventing the sea blockade imposed at Gibraltar by the Phoenicians.

The Greeks named Iberia and set up colonies at Tarraco (Tarragona) and at Saguntum (north of Valencia). In the hope of exploiting the silver deposits of Tartessus, they almost encroached upon the area occupied by the Phoenicians. But the Greeks were a true colonial power, developing the new lands, unlike the Phoenicians who occupied their areas purely as a basis for trade. This great colonization movement of the Greeks, with the resultant spreading of their culture, represented a tremendous achievement in the advancement of mankind.

Meanwhile, political troubles had developed on the Greek mainland. In 521 B.C., King Darius of Persia had become ruler over the greatest empire in the known world, including Egypt, northern India, Mesopotamia, Palestine, Lydia, and Ionia. The Athenians had demonstrated their weakness by being unable to help the Ionian cities defend themselves. So in 492 B.C., Darius sent a large navy to overthrow Thrace and Macedonia. The next attack was launched upon Attica in 480 B.C. A fleet of 600 ships landed but the Persians were badly beaten in the Battle of Marathon.

In defense against further incursions, Themistocles in 483 B.C. used the revenues from the state-owned silver-lead mines at Laurium to build a powerful navy. There was later a terrific battle at Thermopylae where the Spartans fought in defense to the last man against the Persian invaders under King Xerxes. The army of Xerxes then moved into central Greece. A stand was made upon the narrow isthmus of Corinth after Athens had been set afire.

A total of 317 Greek ships, including 130 built by the profits from the Laurium silver-lead mines, soundly defeated the Persian fleet of more than 800 ships, and Xerxes was forced to withdraw. There were no more Persian invasion attempts. One of the major factors in their defeat was the wealth available from the Laurium mines to increase the size of the fleet.

Following the defeat of Xerxes in the Battle of Salamis in 480 B.C., Greece entered its Golden Age of history. Richly decorated temples and statues with marble, ivory, silver, and gold were erected by the famous Greek architects and sculptors. All this was made possible by the abundant deposits of excellent marble and by the wealth won from Laurium.

During the Golden Age, Greece achieved the greatest advancement in intellectual attainments the world had ever seen. But eventually, in 431 B.C., internal political struggles began. By 407 B.C., the treasury was so depleted that gold and silver statues in the Acropolis were melted for coinage. Athens was forced to surrender to the Spartans in 404 B.C. And in 371 B.C., Thebes became the leader of the Greek states.

The mines at Laurium slowly resumed working. But the ore bodies were becoming exhausted after centuries of profitable production. In 103 B.C. the slaves revolted and the mine owners were killed. By the 1st century B.C., only the waste dumps and the slag heaps were being worked. Yet Laurium had been described as "a fountain of silver, a treasure to the land." In a report published in Paris in 1897, it was stated that the old mine dumps contained seven million tons of discarded waste rock, averaging 8 to 10 percent of lead, and 40 to 120 ounces of silver per ton.

A Greek engineer in 1871 estimated that Laurium had produced over ten million tons of lead, which together with its silver content could be worth one billion dollars. At any rate, this was the source of silver and lead (along with gold from Macedonia) that made the famous Greek civilization possible.

This was not the end of the mineral resources of Greece. In 359 B.C., King Philip became the ruler of the Macedonians in northern Greece. In order to build up his army, he took control of the rich gold and silver mines at Mount Pangaeus, the gold-bearing sands of the river Strymon, and the rich silver veins of Krenides. These silver veins had been worked as early as 1300 B.C.

Philip drew 1,000 talents in gold, about six million dollars per year. He amassed a great fortune and hired and trained a professional army, which he led southward to defeat the Thebans and the Athenians. He was now ready to lead the united Greeks against the Persians. But in 336 B.C., he was murdered while celebrating his daughter's wedding. His son, Alexander, inherited his wealth and his army and set out to conquer the world. He became known as Alexander the Great, a brave and handsome king, and a brilliant military and political leader; he was passionately interested in developing the arts and science to spread Greek culture everywhere.

Alexander became master of Greece from one end to the other, crossed the Hellespont (Dardanelles) to Asia Minor, and defeated the Persian army under Darius III. He also liberated Egypt from Persian rule and in 332 B.C. founded the great city of Alexandria. He then moved into and beyond Mesopotamia and captured ancient Susa, worth 250 million dollars in booty; then he advanced to Persepolis and seized the royal treasury of Persia. It took 20,000 mules and 5,000 camels to remove the loot. Alexander caused no damage to Babylon but he destroyed Persepolis unmercifully in a drunken spirit of revenge.

After nine further years of marauding warfare in Persia, Afghanistan, and Baluchistan, he lost 10,000 of his men who died of thirst and exhaustion, and returned to Babylon in 323 B.C. in a poor state of health. He died at the age of 33, and his empire was divided among the generals.

Alexander's reign is called the Hellenistic Age. The immense wealth of gold and silver looted from eastern empires was turned into coinage throughout his realm. The economies flourished. Traders made fortunes and a rich stratum of society developed.

All these conquered lands adopted the Greek language and court customs. Greek arts and letters were everywhere patronized. Buildings and furniture were Greek in design. Greek works of sculpture were in great demand. New cities decorated their temples with Greek statuary. It was difficult for quarries and factories to keep pace with the demand. Greek culture was changing the face of the world.

In mainland Greece, however, there were many internal political struggles. The famous Greek civilization slowly crumbled from within. By 146 B.C., Rome had taken over Greece and Macedonia. A great new empire was to be developed as an inheritance from the Greeks.

The Greek culture had played an incomparable part in developing the world's civilization. Would this have been possible without the wealth derived from Laurium and Pangaeus?

MINING IN THE ROMAN ERA

The classical period of art and literature had been introduced by the Greek civilization dating back to the Achaeans in 1200 B.C. Meanwhile, the Greeks had set up colonies in southern Italy and in Sicily, as well as in southern

France and northeastern Spain. The Greek expansion in the western Mediterranean seems to have been checked only by the Phoenicians who settled in Carthage in 814 B.C., as well as in many other Mediterranean ports, including Cadiz in Spain.

But now a third Mediterranean power began to develop in 800 B.C. Exiled because of a long famine, the Lydians migrated to northern Italy and settled in Tuscany (Etruria). Here they found a fertile district that was also rich in natural resources, such as deposits of iron and copper ores, with some tin (see figure 9.1). With their expertise developed earlier in Lydia, the Etruscans were able to capitalize on these ore deposits and produce magnificent articles, ornaments, and statuary in bronze, alabaster, gold, and silver. Their superior armor and weapons were made of bronze and iron, exquisitely chased and engraved. Etruscan bronzes are famous treasures in the museums of the world of today.

The Etruscans expanded their metal industries and developed a significant maritime trade throughout the Mediterranean. They imported many art forms from Greece and Asia Minor. The development of chariots, together with their superior armor and their fearless nature, gave them considerable military supremacy.

They became allies of the Phoenicians and tried to drive the Greeks out of southern Italy, but failed in 540 and again in 524 B.C. However, they captured Rome, and Etruscan kings ruled Rome from 616 to 509 B.C. with an assembly of citizen warriors and a senate of aristocrats. Later they were invaded by the Gallic barbarians from the north who conquered Etruria and their power as a nation came to an end.

Meanwhile, Latin tribes had settled around the Tiber River in Rome. They had no skills or artistic background but were crude, staunch, and patriotic barbaric tribes. Yet they were able in time to establish an empire of tremendous power and status. Here the great Roman Republic was formed in 509 B.C., following the defeat of the Etruscans. The Romans gradually extended their influence over neighboring regions and absorbed their cultures. By 270 B.C., the Republic covered all Italy south of the River Po.

To gain full command of the Mediterranean, however, the Romans needed to subdue the Phoenicians. The First Punic War was begun in Sicily in 264 B.C. By 241 B.C., they had overrun Sicily and Sardinia and had extracted an indemnity of 3,200 talents of silver (about five million dollars) from the Carthaginians (Phoenicians).

But Carthage, although richer than Rome, was hard-pressed to meet these payments. So Hamilcar turned to Spain for relief. The new city of Cartagena was founded in 225 B.C. and soon became prosperous, based upon the wealth derived from gold, silver, and lead in the Tartessus hinterland.

Following the death of Hamilcar in 229 B.C. and the assassination of his successor Hasdrubal, Hannibal became the military commander of the Carthaginians at the age of 25. In 219 B.C., he captured Saguntum and thereby ushered in the Second Punic War (219 to 201 B.C.). He then led 50,000 infantry and 9,000 cavalry over the Pyrenees and crossed the Alps into Italy without opposition. The Romans were thrown into disorder.

But Hannibal was unable to take Rome. In fact, he was obliged to return to Africa to defend Carthage against an attack by Scipio Africanus. The Second Punic War ended with the defeat of Hannibal. As a result, the Carthaginians had to surrender all their Mediterranean islands and also their Spanish possessions. The real prize won by the Romans was the mineral wealth of Spain. This was a tremendous factor in the development of the Roman Republic.

Later, as Carthage was reviving due to a resurgence of trade, the Roman censor, Marcus Porcius Cato, visited Carthage and returned to Rome, leading an outcry: "Carthage must be destroyed." Rome attacked in 149 B.C. and besieged Carthage in the Third Punic War. Of 500,000 people, only 50,000 survived, and they were sold as slaves. According to reports, the ancient city was "burned, ploughed up, seeded with salt, cursed and made unfit for habitation."

Meanwhile, King Philip of Macedonia had earlier sought to become an ally of Hannibal. Macedonia was overcome finally in 168 B.C. and the Romans gained a tremendous amount of booty, just as they had when Scipio defeated the King of Syria, gaining the mineral lands of Anatolia.

Moreover, the Macedonians had forewarned the Greek states about the growing power of Rome, but these Greek city-states had their own internal problems and were powerless to defend their liberty. Thus, Rome took Corinth in 146 B.C. and finally overran the whole of Greece. The treasure chests of Greece were transferred to the vaults of Rome. But all was not lost: the cultural heritage of Greece was passed on to the world through the Romans.

Now the Romans turned to their own newly acquired lands. Spain was the most highly mineralized area in the known world. She was a key acquisition for the Republic. With their great organizing skill, solidarity, and statesmanship, the Romans settled, civilized, and developed their barbaric lands.

Rome first needed a reliable currency to finance her expansion efforts, to stimulate her trade, and to build up her military strength. The governors of new provinces and the generals were empowered to establish local mints to coin money to pay the troops. This enhanced the morale and loyalty of local garrisons. To expand her currency, Rome was able to turn to Spain for silver. New supplies of other metals were also needed for munitions and to feed the growing craft industries. The various tribes of Spain were brought under control. The silver, copper, and gold deposits, earlier worked by the Greeks or Phoenicians, were now revived.

For 200 miles across southern Spain from east to west, mines dotted the landscape. This region was the greatest mineral province ever discovered. At the eastern end of the Sierra Morena, the Baebalo mine alone had furnished Hannibal with 300 pounds of silver a day.

In 208 B.C., the elder Scipio defeated the Carthaginians and took possession of this region. The Romans worked it intensively. Some ore bodies in this area were worked 600 feet deep by shafts, and drifts 3,000 feet long were driven into the ore, using fire-setting methods.

Some of these mines were rich enough to warrant reopening in recent times. Some interesting relics found were the Archimedean screws (cochleas) used to lift water, described in chapter 14.

For several centuries, Rome drew its great wealth from the lead and silver of Cartagena and Almeria, the mercury mines at Sisapon (now Almaden), the famous copper and iron ores of Rio Tinto, the gold of the Tagus River, and the rich tin deposits of Galicia (see figure 9.1). The conquest of Spain and the fabulous booty seized throughout the east brought untold wealth to the Roman state and her citizens. In about 60 B.C., Julius Caesar rose to power as consul of Rome and commander of her armies, after having acquired a great personal fortune during his campaigns in Spain. His next plan was to move into Gaul to seek still more treasure. Many local barbarian tribes had to be subdued, yet there were very few rich ore deposits in Gaul. Moving further north, he encountered the Veneti tribe who had held a monopoly over the rich tin maritime trade between Cornwall and the mouth of the river Loire. They proved to be staunch defenders, and until their subjugation, they refused to disclose the source of their tin.

Actually, the alluvial tin oxide from the streams of Cornwall was smelted with charcoal. The tin ingots were warehoused at Ictis, a Cornish offshore island, now known as St. Michael's Mount. It was the Veneti who had previously supplied the Greeks and traded their tin to the Phoenicians from the Cassiterides Islands near the mouth of the river Loire. Caesar's army had to exert superhuman efforts to overcome the Veneti.

The disclosure of the source of tin in Cornwall and the need for lead inspired Caesar to invade Britain. But first he had to return homeward to repel a number of invasions of Gallic and Germanic tribes.

Then, in 55 B.C. Caesar made a brief but unsuccessful attempt to invade Britain. He landed on the east coast but had to withdraw to put down further revolts in Gaul. In 53 B.C., he overcame all the Gallic tribes but by now he had troubles at home. He returned and defeated his rival Pompey, then went to Egypt for an amorous alliance with Cleopatra, and later to Anatolia to defeat King Pharnaces. In 45 B.C., he returned to Rome with a tremendous booty and was accorded a triumphal entry; he celebrated lavishly with great gifts to his soldiers and favorites, and spent large sums on public festivals and celebrations. Caesar was elected consul for ten years and dictator for life. But he was assassinated five months later, and the Roman Republic was ended.

Caesar had introduced many useful reforms: he reduced debts, and spent heavily on public improvements, including a vast system of roads; he granted citizenship to the provinces and thereby consolidated the nation.

Following 20 years of civil war, Octavian (Augustus) came to rule as the first Roman emperor, but he retained many of the democratic principles of the Republic. This was the beginning of the Roman Empire. Two hundred years of peace followed in what was known as the Pax Romana, the Golden Age of Rome. Rome became the most beautiful city in the world. An extensive program of public building was commenced, mostly in marble. Roads, aqueducts, fountains, and public baths were built. The coinage system was also expanded to extend further the trade throughout the known world, to finance new conquests, and to maintain the garrisons.

Based upon the expertise developed earlier in Mesopotamia, Egypt, Anatolia, Tyre, and Greece, magnificent works in pottery and glass were produced, as were famous bronze and silver wares. The considerable revenues from mineral production in Spain and other provinces were the basis for Rome's military and trade supremacy and for the affluence and opulence of her society.

In support of this movement, the political policy was designed to take over new areas that had significant mineral potential. In these provinces, roads were extended and colonial settlement was accelerated. It was just as true then as today: the nations that sponsored mineral production activities became the more affluent.

Under Claudius in 43 A.D., Britain was again invaded. Despite the fierce resistance of the native tribes, the Romans were able to find gold and iron deposits, but more particularly lead in the Mendip Hills and in Derbyshire (see figure 10.1). Many old lead smelters have since been found, with ingots bearing the names of emperors and the corresponding dates. At least 70 lead ingots, weighing from 125 to 230 pounds have been recovered. But the Romans were not the first of the lead miners in these areas. Some Celtic artifacts have been recovered, showing that the lead mines were worked centuries

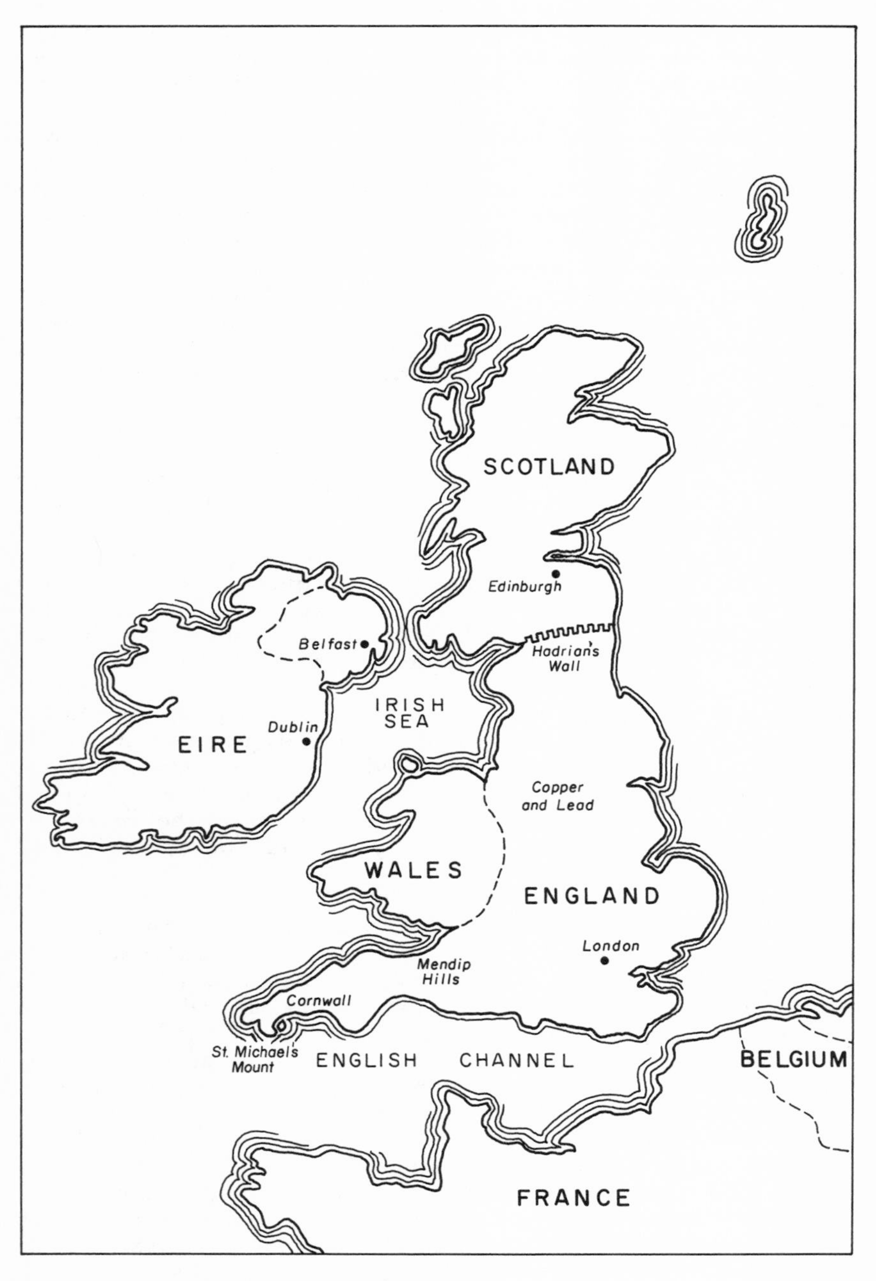

Fig. 10.1. Map of Great Britain.

earlier. However, the Romans applied their customary efficiency and organization to the operation of these old mines.

The main prize won by the invasion of Britain was the acquisition of large supplies of lead. Throughout all their provinces, the Romans made good use of lead for roofing public buildings, for lead piping for the water supply and drainage systems of their cities, and for ornately engraved lead coffins. Pewter (an alloy of tin and lead) became popular for drinking vessels, serving dishes, and for dinnerware.

The lead ores in Britain were not only more abundant, but were more easily mined than the deeper-seated lead veins in Spain; considerable amounts of silver were also recovered by cupellation. Lead was mined by the Romans in Flintshire, Derbyshire, Yorkshire, Shropshire, in the Mendip Hills, and at Alston Moor, near Hadrian's Wall. The lead and silver mined in Britain were now more important to the Romans than the more difficult deposits in Spain. As the tin mines in Spain and Portugal became exhausted, the Romans turned to Cornwall.

Iron ore deposits in Britain were also developed by the Romans. Many of the oak forests were denuded in order to produce charcoal to smelt these ores. The Romans exported iron to the continent. Centuries later, Britain was to initiate the Industrial Revolution, based mainly upon her ability to produce iron and coal.

At first the Romans were slow to develop the tin mines of Cornwall, worked earlier by the Veneti. By this time, Rome had subdued the tribes in the northwest of Spain, and the tin mines of Galicia were being energetically exploited. But in the 3rd century A.D., mining was revived in the streambeds following the extension of the road system to Cornwall. In this area, a considerable output of tin was produced until 410 A.D. when the Romans had to withdraw their forces from Britain to defend the homeland against barbarian attackers. Following their withdrawal, Britain was invaded by Angles, Saxons, and Jutes.

However, the effect of 400 years of Roman influence had brought Britain from a barbaric age to a civilized community, founded upon the Roman system of law and order. A sound system of government, commerce, law, and education had developed, based upon the Latin language. The Romans left behind great walled cities, municipal baths, public buildings, durable stone highways, and quarries and mines that were to become the foundation of trade, commerce, agriculture, and culture in Britain.

Other areas of interest to the Romans, because of their potential mineral wealth, were Noricum (Austria), Dalmatia (now in Yugoslavia), and later, Dacia (Romania).

As early as 130 B.C., Gaius Marius returned from an expedition to northern Italy with rich booty in the form of gold taken from the Cimbrian tribes. Later, a law was enacted prohibiting the employment of more than 5,000 workmen in the streambeds from which this gold was taken. It was feared that the gold output would become so great that its value would decrease, as had happened in 170 B.C. at Aquileia, 25 miles north of Trieste.

But the greatest gold discovery in Roman times was in Dacia, in the Transylvanian Alps (in present-day Romania). Marcus Trajan subdued the Dacians in the early years of the 2nd century A.D. and captured 10,000 prisoners, 500 tons of silver, and 250 tons of gold. When this booty reached Rome, it brought about a gold rush to Dacia. This new province was systematically colonized. Substantial roads and towns and great walls around the cities were built to keep out neighboring barbarians.

Trajan's triumphal return to Rome was accompanied by the distribution of 75 million dollars to 300,000 citizens. With the rest of the booty, a magnificent new forum was built. This included Trajan's Column, erected in 114 A.D. to commemorate the conquest of Dacia. It is 97 feet high, superbly carved in marble in an ascending spiral of reliefs displaying an authentic record of the military victory, and symbolizing nearly 400 years of Roman supremacy. Trajan's ashes are deposited in a golden urn at the base. This magnificent column may still be seen in Rome.

Hadrian was one of the most capable emperors of Rome. He set out to make Rome the greatest city ever known, in tune with his impressions of the cultural heritage of Athens.

The *Pax Romana* flourished during the reigns of Trajan, Hadrian, Antoninus, and Aurelius. This period was the "most happy and prosperous the human race ever enjoyed in the history of the world." Most of the prosperity was derived from the mineral resources of Spain, Britain, and Austria, whose riches sustained the economic and cultural wealth, as well as the power, of the Roman Empire.

For over 400 years the Romans owned all the profitable mineral resources of the known world. The vast wealth of the Roman treasury developed a sense of exhilaration, excitement, and activity among the officials, bankers, merchants, and citizens. The city of a million people was resplendent with the grand marble forums and palaces, the Colosseum, the Circus Maximus, and several amphitheaters built from marble to provide chariot races, gladiatorial contests, and other forms of entertainment for the people.

The homes of the middle class and the resplendent villas of the wealthy were richly adorned with sculptures, paintings, and classical objects of art imported from abroad. Villas and public buildings were faced with Grecian columns in marble and in granite; floors were of marble, onyx, or alabaster. Bathroom fittings were of silver and sometimes of solid gold.

Articles of gold jewelry were richly set with precious gems from the East. With the advent of the glass-blowing art from Tyre, such ornaments as vases, cups, phials, and bowls of colored glass were eagerly collected. Painting of murals and friezes was highly regarded.

And, as in all civilizations where affluence and a luxurious standard of living prevail, the arts became intensively cultivated. But sooner or later, luxurious pleasure and self-indulgence promote laziness and moral laxity.

After 180 A.D. and following 200 years of the *Pax Romana*, Rome began to decay. A prolific rate of government spending, imbalance of foreign trade, and corruption brought on several waves of currency devaluation. The gold and silver currency was debased by alloys of zinc, silver, copper, and lead.

By 218 A.D., Rome faced economic disaster. The provinces were overrun with Germanic tribes. Her rich storehouses of mineral wealth, which had sustained her opulence for nearly 400 years, were no longer available. Luxury had given way to indolence, and now to political apathy. Rome collapsed from within. And now she became a simple target for invasion. The Ostrogoths sacked Rome and the Vandals and others overran Gaul and Spain. Tin mining was brought to an end in Spain in the 3rd century A.D. In 455 A.D., the Vandals ruthlessly "vandalized" Rome by wanton destruction. Rome's domination as a world power had come to an end. Yet Rome's gifts and treasures for mankind were beyond description.

Now the civilization of the world was to become eclipsed by the Dark Ages, a blank period of six centuries.

Although the Roman Empire had faded into obscurity, her glorious heritage lives on. It survived the Dark Ages and now, 2,000 years later, Rome can justly be termed "The Eternal City."

MINERAL DEVELOPMENTS IN THE ORIENT

Not much is known about mining and the use of minerals in Asia during this period. Figure 4.14 shows some of the more important mining centers. Most oriental countries based their economies upon agriculture; minerals were used mainly in the ceramic arts, including the early use of coal and coal gas for firing their ceramic furnaces. Precious metals were used primarily for currency purposes, and salt was required as a food supplement.

Some copper was mined in China at Shuo Shan in the Shansi province in the 3rd millennium B.C. Presumably this copper was used for bronze and brass production. The Chao

Yeh gold mine in the Shantung province was opened up in 1007 A.D. It is currently producing gold, copper, silver, and sulphur from a daily production of 500 metric tons of ore.

The mining and smelting of iron ores in China have been known since 800 B.C., especially in Shansi. Agricultural implements have been made of iron since that time and iron chains for suspension bridges began being produced a few centuries later.

In India and Japan, steel was first made by the slow heating of wrought iron with charcoal and sawdust in closed crucibles. In fact, the prized sword blades of Damascus were forged by blacksmiths using imported Indian steel. The Romans regarded seric iron as the best steel available. It was generally believed to be of Chinese origin, but it actually came from India, via Abyssinia. In Delhi, the Iron Pillar may still be seen near the Qutb Minar minaret. This is a solid column of unrusted wrought iron erected in 310 A.D. by the Rajah Dhara.

Both iron and salt became a national monopoly in China in 119 B.C. Iron coins were minted around 525 A.D. In 914 A.D. the government monopoly over iron production was removed, but it was again revived in 1083 A.D.

The Great Wall of China, extending for more than 1,400 miles along her northern boundary as a protection against foreign invaders, was built in the 3rd century B.C. Gunpowder was invented in China during the Wei Dynasty in about 250 A.D.

Otherwise, perhaps because of Confucian teachings emphasizing the need for a simple life, mining was discouraged and even prohibited during several of the Chinese dynastic periods up to 1650 A.D. At other times, it was actively encouraged.

The Zawar mines, 28 miles south of Udaipur, India, were rediscovered in about 1400 A.D. These mines produced silver, tin, copper, lead, and antimony.

The famous Bawdwin Mine containing silver, lead, and zinc was opened in Burma in 1412 A.D. It is still in operation, although production has fallen off since it became nationalized after World War II. The immense golden dome of the Shwedagon Pagoda may be seen at Rangoon.

The Far East is known to have been one of the main sources of the tin needed for bronze production in the Middle East. Tin was regularly traded along the main eastern trade route during the Bronze Age. Stream tin is still produced in significant quantities in Thailand, Malaysia, and Indonesia.

The Kamaishi Mine in northeastern Honshu is the oldest in Japan; it was first cited in 1822 and became more prominent in 1857 when Japan's first blast furnace was blown in. The mine produced iron ore concentrates of good quality. Copper ore bodies were later discovered in the vicinity.

One of the substantial branches of the mineral industry for which the Orient has become famous is the gem industry. India has produced diamonds and sapphires since about 700 B.C. Sri Lanka (Ceylon) has become known as the "Isle of Gems." Jade has been mined in China since 1100 B.C., and also in Burma. An effigy of Buddha, carved out of jasper (supposedly emerald), and another cast in solid gold and weighing five tons, may be seen at Bangkok. Others have been cast in bronze, such as the Daibutsu, cast in 1252 A.D. at Kamakura, Japan; it weighs 122 tons and stands more than 44 feet high.

Another bronze statue of Buddha in the Todaiji Temple at Nara, Japan, stands 53 feet high and weighs 250 tons. It is the world's largest bronze image of Buddha. How were they able to make such enormous castings in those days?

Many mineral substances such as iron ochers, lead oxides, antimony sulphide, and mercury salts were mined and used generally for cosmetic purposes, and also as decorative pigments.

THE MIDDLE AGES (MEDIEVAL PERIOD)

The medieval period is generally supposed to cover the years between 450 and 1450 A.D. The first part, following the classical age and the fall of Rome, from 450 to 1050, is also known as the Dark Ages.

During the Roman era, the Germanic tribes north of the Danube and east of the Rhine were constantly threatening Gaul. Among them were the Ostrogoths who sacked Rome in 410 A.D. and the Vandals who completed its destruction in the year 455. The provinces were similarly burned and pillaged by these barbarians. As a result, the local peasants had to seek the protection of the lords or religious leaders, who set up new villages clustered around a manor or monastery. In order to have their security guaranteed, these peasants had no recourse but to yield to a bond of serfdom to their overlords for life. From this state of affairs, feudalism and the manorial system developed (800 to 1250 A.D.). In the Middle Ages, feudalism became the only organized living code in Europe, except for Spain where the Moorish regime maintained a high level of cultural affluence based on Córdoba, the most impressive city in Europe at that time.

In practice, each manor or monastery had to become self-sufficient, especially in food production, and life was reduced to the use of the local resources in that area. Trade was stifled. The mining of minerals became severely restricted, except for iron necessary to make weapons of defense and tools for cultivation. Interest in mining was soon lost and mine workings became overgrown with natural plant life.

However, in East Africa, the Iron Age races had developed some skill in metalworking. It is believed that gold smelting was known to the peoples of Zimbabwe, a civilization that existed between 700 and 1500 A.D. Early African metalworkers also used the Katanga copper cross as a primitive form of currency.

As the existing supply of Roman currency became depleted, and as mining came to a virtual standstill, Europe began to stagnate. Salt mining continued, however, and alluvial gold washing was furtively carried on by individuals. Any building stone that was required was garnered by the robbing of existing structures. Nevertheless, certain areas of mineral production never completely lost their fire. Sporadic mining still continued, but at a slow pace, in Saxony and Bohemia, and in Derbyshire and Cornwall.

In 486 A.D, Paris had been set up as the capital city of the Franks under Clovis. The Moors had invaded Spain in 711 A.D. and were threatening Gaul, but they were defeated by Charles Martel in 732. This action introduced a new dynasty to Europe, the Carolingians. Later, the grandson of Martel, Charles the Great (Charlemagne) created the opportunity for Europe to emerge from the Dark Ages.

In 785, Charlemagne subdued the Saxons on his northeast border and the Avars in Hungary. He then annexed Bavaria and set up fortresses on the frontiers of Spain, Italy, Bohemia, and Thuringia. He now had most of Europe under effective control.

In 800, Charlemagne was crowned Emperor of the Romans by Pope Leo III. In laying the foundation of the Holy Roman Empire, he became the first powerful sovereign in western Europe, using and stimulating Christianity as a unifying force. Charlemagne established magnificent churches, monasteries, and bishoprics across the land. In particular, he built a beautiful basilica in Aachen and adorned it with gold and silver, using rails and doors of solid brass. Marble pieces were introduced from Rome and Ravenna. This magnificent church still exists in Aachen in a fine state of preservation.

The Carolingian period proved to be a turning point in history. While promoting the Christian faith, the monks were able to study and preserve many priceless works of Latin literature and to pass on the scientific and artistic legacies of former civilizations. This intellectual revival became known as the "Carolingian Renaissance."

A new spirit had arisen. Goldsmiths and other metalworkers were inspired to show their skill. Beautiful jewelry, made from bronze and gold, set with gems, was created. The demand for metals increased, and particularly for lead, for roofing the new cathedrals. Skilled workers from Saxony had worked the mines at Kremnitz in the 6th century and those at Schemnitz during the 7th century.

Charlemagne extended his influence beyond the Rhine and the Elbe. In the year 770, he revived the rich silver and gold mines of Schemnitz and Kremnitz, actually about 80 miles northeast of Bratislava, in present-day Czechoslovakia. The Amalia Mine near Schemnitz was later worked to a depth of 1,800 feet. These mines produced large amounts of silver which was now used instead of gold for coinage. The reopening of these mines was the starting signal for further mineral exploration. As precious metal production increased, there was renewed prosperity, bringing Europe out of the Dark Ages.

But as fast as Charlemagne distributed control over these mining areas to his sons, they in turn transferred them to relatives and friends and members of the nobility for favors received. In time, in the absence of a central power after the death of Charlemagne, this practice grew into organized feudalism, in which each lord or master owned the useful land (and the mineral rights) and used his serfs to work it in exchange for his patronage and protection.

Later, these great landed estates grew in importance and became duchies throughout France and Germany; each became a self-contained seat of power. There were many struggles among them to gain the leadership of Europe. These duchies later became provinces of either France or Germany.

In 919, the dukes of Germany recognized the Ottonians as the supreme authority over the land east of the Rhine. Otto the Great was a progressive leader. He built fortresses along the borders and encouraged colonists to establish towns nearby, and to explore and develop virgin lands. But during the 10th century, France and Germany badly needed a supply of silver to expand their currencies and trade, and were also in need of copper, lead, and tin to support the growth of their metalworking industries.

One of the major finds (in 938) was on a forested hill in the Harz Mountains of lower Saxony. Today this is the famous Rammelsberg Mine, the only mine worked continuously for over 1,000 years (from 968 A.D.). Rammelsberg soon became the chief producer of copper, silver, and lead in central Europe. This mine has provided valuable economic assistance to Germany for over a thousand years. It is situated near Goslar (see figure 7.1 and plate 10.1).

The mining of lead ore in Britain was begun by the Romans. Derbyshire, a lead belt, 30 miles by 12 miles, produced lead for many centuries (see figure 10.1). Lead deposits were also worked in Shropshire, Cornwall, Wales, Scotland, Ireland, the Isle of Man, and the Mendip Hills.

The first war of the Crusades began in 1096 as a holy war to recapture Jerusalem from the Muslims. These wars had a tremendous impact upon European life. Among other advantages, they reestablished the desire for trade with west-

Plate 10.1. The Rammelsberg Mine.

ern Asia. Western Europe developed a considerable interest and demand for imports of quality from the Middle East. The Mediterranean seaports, especially Genoa and Venice, began to gain a great advantage from this trade. The long dormant spirit of western Europe began to flicker with the desire for goods following the development of the Rammelsberg Mine and the trading potentialities of the Middle East. But a shortage of western currency had brought about the need for a renewal of the barter system.

In order to support the costs of the Crusades, many feudal lords had to divest themselves of some of their lands. This gave rise to the development of independent towns and the need for some free workmen. In addition, many serfs were encouraged to find mineral deposits for the duchies in exchange for their freedom. One of the few opportunities open to a serf was to become a prospector and find a mineral deposit, stake his claim, obtain a concession or license from and pay a royalty to his landlord, and work his property. In this way, he could become free from serfdom, become his own master, and perhaps become wealthy. Such freemen gradually developed the hardy and resourceful spirit of prospectors and miners, and collectively played a considerable part in the emergence of today's civilized world.

During the 12th century, the revival of civilization, and the need to finance the Crusades, created a burgeoning demand for the supply of metals, particularly silver for coinage. The finding of Rammelsberg stimulated prospecting activity in Germany and elsewhere for this purpose. Mining had been firmly established at Wildemann (1225), at Zellerfeld (1243), and at Clausthal (1251) in the Harz Mountains. The lead, silver, and copper mines in southern Spain, worked by the Romans, continued during the Moorish regime, but production tapered off soon after they were expelled in 1492. The local mines were neglected as rich deposits became available later in Spanish America. Nevertheless, the famous mercury mining center at Almaden continued to produce.

Near the headwaters of the river Elbe, in the Erzgebirge, the next important mineral find was to occur in 1170 A.D., when a silver rush developed. On the western slope of the Erzgebirge, in upper Saxony, in the granite hills around Freiberg, Schneeberg, Zinnwald, and Altenberg, veins of tin, molybdenum, tungsten, silver, bismuth, and cobalt were later found. Further west, there were silver, copper, and nickel deposits (see figure 7.1). On the eastern side, in Bohemia, rich seams of native silver and other silver minerals were associated with bismuth and cobalt. These surface signs developed into deep mines at Joachimsthal (now Jachymov).

The Erzgebirge became famous all over the world for its wealth and the range and variety of its mineral output. It developed into one of the great metallogenetic provinces of the world. Here they coined silver _thalers_ (dollars).

Many more finds occurred in what is now known as Czechoslovakia; these were in mineral belts both west and east of Prague. The old Bohemian town of Iglau was one of the first mining towns ever to be granted a charter, in 1249. From this concept, one of the first practical codes of mining law developed. Bohemia became the wealthiest of all countries in late medieval times.

Prospecting continued further into Hungary and into the southern Carpathians of Romania. The mines worked once by Charlemagne in Schemnitz and Kremnitz in the Slovakian Ore Mountains were reopened, and more gold, silver, and copper were produced. The miners of Saxony had earned such a good reputation that the merchants of Venice engaged them to develop mines in Serbia. This brought about the development of Dubrovnik as a port city on the Dalmatian coast. In addition, gold, lead, silver, and copper mines were worked during this period in the Austrian Alps. The famous Stora Copper Mine in Falun, Sweden, was in operation before 1288 A.D.

And so, during the 12th and 13th centuries, following the discovery and development of the mineral potential of central Europe and Bohemia, Europe began to experience a brilliant cultural period. Serfs became free. Frontier villages grew into cities. The old trade routes became reestablished. Metalworking industries received a great boost. Small traders became merchants and bankers, and a new class of businessmen developed. They were known as burgers in Germany, burgesses in England, and the bourgeois in France. It was the adventurous spirit of the prospectors, supported by the burgers, that led Europe out of the Dark Ages. And with it all, Christianity developed as a unifying force. To support this religious trend, a strong movement to build lavish churches and cathedrals began. The medieval metalworkers, sculptors, painters, and stonemasons developed architectural edifices of grandeur in the shape of magnificent Gothic cathedrals. This movement had been inspired by Charlemagne, who built his famous cathedral in Aachen in the year 800. During the 11th century, over 1,500 churches in the Romanesque style were built in France alone. By 1250, there were over 500 Gothic cathedrals in France and many in Germany.

The Gothic designs incorporated ornamental stone carvings and bronze sculptures, with tall majestic spires and belfries. They were embellished with all manner of Biblical figures and religious scenes, with heavy bronze doors, ornately sculptured and chased, and bronze altars decorated with gold and silver. Most of these metals came from Bohemia.

In similar fashion, bronze metal was demanded in great quantities for church bells all over Europe; and later the bell-founding techniques were used for the casting of cannon barrels. With these new developments in warfare, England was able to defeat the French in the Battle of Crécy in 1346.

As time moved on, the mining of deeper deposits and complex ores brought their problems, and more easily worked silver mines were becoming exhausted. Then in about 1350, the Black Death took one-third of the population and seriously reduced the work force. Many internal political struggles between dukedoms and states also seriously affected mineral production, and opposing military forces destroyed many of the mines and smelters. Thus, after 200 years of regular production, a depression set in and the mining of metals, except for iron needed for warfare, came to a halt. Iron was necessary for the fabrication of armor, sword blades, and the steel crossbow. Europe could not keep its economies running smoothly in the face of such political problems as the Hundred Years' War (1337-1453) and the Hussite Wars (1415-1435).

So, in the face of these problems, the mining industry found it necessary to introduce and invent improved methods. Water wheels were installed for operating machinery, following their earlier use by the Chinese, Moors, and Romans. Where water was not available, or where slave power was scarce, animal power was pressed into service. Metallurgical problems also arose. It was difficult to separate certain minerals from complex ores. Solutions to all these problems called for a great deal of capital financing to install expensive plants, introduce new methods, and develop the mines at greater depths.

In order to secure the necessary capital funds, miners had to sell shares to merchants, monasteries, and town councils. Sometimes they borrowed from bankers and traders. When they ran into debt, the creditors gained control and engaged experts to run the mines, employing the miners on a wage basis. This brought about an era of capitalistic control which was necessary to stimulate invention and to develop new techniques. Much of the capital was provided by wealthy financiers. Even kings depended on the power of these financial giants. Some of them came into possession of large mineral empires in the various European states.

In the metallurgical field, many complex ores of copper could not be profitably mined because they contained too much silver. But in 1451, Johannes Funcken developed the age-old method of separating the silver from copper by liquation with lead and by subsequent cupellation of the lead-silver product. As a result, both silver and copper could be separated and sold, and the lead recovered, thereby enhancing the revenue from such mines. Whereas formerly these ores were unprofitable, now they became highly profitable. This innovation provided a tremendous impetus to European mining in the 15th century. Another important development at this time was the invention of larger furnaces for smelting iron ores. Three new types were introduced in the 15th century in the Harz Mountains area. The additional output obtained from a single furnace greatly reduced the cost of smelting iron ores.

And so, from 1450, metal mining in Europe boomed and the industrial growth of Europe expanded. The technical developments in Germany spread to France, England, Scandinavia, and Spain. The production of silver in Germany increased fivefold between 1460 and 1550. The Fugger family of merchant bankers in Augsburg became the wealthiest in Europe.

Silver deposits were discovered in St. Andreasberg in 1480 and mining began in 1528. These famous lodes averaged 6 percent lead, 2 percent zinc, 1 percent copper, 1 percent antimony and arsenic, and 1 percent cobalt and nickel, with some gold in gangue minerals of quartz, calcite, barite, and siderite.

In 1623, the Kongsberg silver deposit in the south of Norway was discovered. It became the richest silver mine in Europe.

Many technological methods and devices were developed in the Harz Mountains during this period. Extensive reservoirs were built to supply water-powered devices. Until 1840 these mines were noted for their technical advancement in operating methods. Many other countries, including England, learned their mining skills in this district. Hard-rock mining expertise was developed in Cornwall from lessons learned in the Harz Mountains, and subsequently passed on to North America, Australia, and Africa through the ubiquitous Cornishmen. Queen Elizabeth in 1562 secured the services of 4,000 German miners to mine and smelt copper ores at Keswick, in the English Lakes District.

In 1556, Georg Bauer (known as "Agricola") published a most comprehensive outline of mining and metallurgical practices of the day. This book was written in Latin under the title of De Re Metallica. It represents the fruits of a life experience in mining, and depicts the methods used in the most advanced areas during the 16th century, principally in Saxony and Bohemia. This book was translated into English in 1912 by Herbert C. Hoover and his wife, Lou Henry Hoover. Their translation is regarded as a classic treatise of great integrity. Hoover was a celebrated mining engineer.

The mines of Cornwall also enjoyed success in this boom period. Tin had been mined in Cornwall from streambeds since 1000 B.C., possibly by the Veneti. The Phoenicians, Greeks, and Romans in their turn depended on Cornwall's production. But for several centuries after the Roman withdrawal, mining activity had waned.

Nevertheless, mining in Cornwall revived, especially following the Norman invasion of 1066. Much of Cornwall's copper and tin was used for making pewter ware, cannon barrels, church bells, and for copper plating.

King Richard the Lion-Hearted, who was imprisoned by Duke Leopold of Austria in Durnstein Castle, was eventually released in 1194 by a ransom gathered in England, mainly from Cornwall and also from the silver-lead mines.

In the year 1201, King John granted the tin mining districts, called the "Stannaries," a Royal Charter. This set out the rules for mining and smelting tin ore and for obtaining fuel from nearby forests. It also specified the royalty to be paid. This charter released the tin miners from serfdom. In 1337, tin production from placer deposits reached its zenith. The cassiterite was finally traced to the veins occurring in the pink granite country rock. By 1450, the miners were prospecting for these veins and were opening new mines. After the 17th century, nearly all the tin produced was won from veins.

THE RENAISSANCE PERIOD (1350-1550 A.D.)

The mining and industrial booms of the 15th century stimulated the European economies and prosperity abounded. These countries emerged from a long period of stagnation. Modern and innovative trends of thought developed. This is known as the "Renaissance Period." It represented the reincarnation of cultural achievement following its dormancy during the Dark Ages. The Renaissance developed first in Italy, primarily in Florence, where for many years there had smoldered a deep abiding interest in the earlier classical cultures of Greece and Rome.

Florence had become a flourishing, prosperous city. It was situated midway between the two great trading ports of Genoa and Venice, in a strategic position to profit by and to be inspired by the wares of cultural value coming from the Middle East and Central Asia. Its merchants and its banking families had a deep cultural interest in reviving the classical arts. They sponsored the employment of talented artists and craftsmen to embellish their places and churches. Many of the fine villas and estates became treasure houses of classical art. The interest in the revival of learning, literature, art, architecture, and humanism received a sound impetus.

Nevertheless, it is not possible to imagine how the Renaissance period could have developed so strongly except through the discovery of rich mineral deposits and the corresponding urge for exploration and development. And, with an increasing supply of metals and minerals, the craftsmen and artisans of the day were strongly supported by the rich merchants who craved for the ornate and extravagant embellishment of their palaces and churches. New thoughts and ideas were encouraged by the opportunity for a creative spirit of endeavor. The Renaissance period led to an "Age of Discovery."

People had heard about the luxuries of the Orient during the wars of the Crusades. They were also bewitched by the tales of Marco Polo, a Venetian who travelled to China and returned in 1295 to relate his vivid experiences in Mongolia at the court of Emperor Kubla Khan, whose guest he was for 17 years during a journey of 25 years. Others had seen ships laden with exotic treasure in the ports of Genoa and Venice. Adventurers were thereby inspired to seek the source of this wealth.

Several Portuguese explorers had rounded the Cape of Good Hope and reached India. Portugal, Spain, and Italy derived much of their gold for coinage during the 14th and 15th centuries from West Africa, typically from the Gold Coast (of Guinea).

Then Christopher Columbus set out from Spain on a journey westward to seek the treasures of the Orient. On the way (October 12, 1492) he made a landfall on the Bahaman island of Guanahani, now known as Watling. He had actually rediscovered the American continent. Centuries earlier, the Vikings, the Irish, and the Phoenicians had visited this continent, but had not sustained their discoveries.

THE CONQUISTADORES

It was the hope of finding fabulous treasure in Cipangu, Japan, (and in Cathay, China) that inspired Columbus to venture westward, following the glowing descriptions of great wealth related by Marco Polo. His expedition was sponsored by Queen Isabella of Spain. He was to receive ten percent of all the precious metals and gems discovered and was to be appointed governor-general over the lands annexed.

However, after landing at Guanahani, there was some disappointment to find that the natives wore no golden jewelry and knew nothing about the existence of Cipangu. So the expedition moved in turn to other islands, including Cuba. Tropical plants and flowers grew in profusion, but there was no sign of golden cities or treasure (see figure 10.2).

By a stroke of fate, the expedition was later shipwrecked on the northwest coast of Hispaniola, in an area now ruled by Haiti. Here, the native women wore gold necklaces and bracelets. It so happened that this was the only Caribbean island that yielded gold. Members of the expedition later found gold in one of the streams.

A settlement was made here, the first developed by Europeans in the western hemisphere. Columbus hastily returned to Spain to report the good news. He later estab-

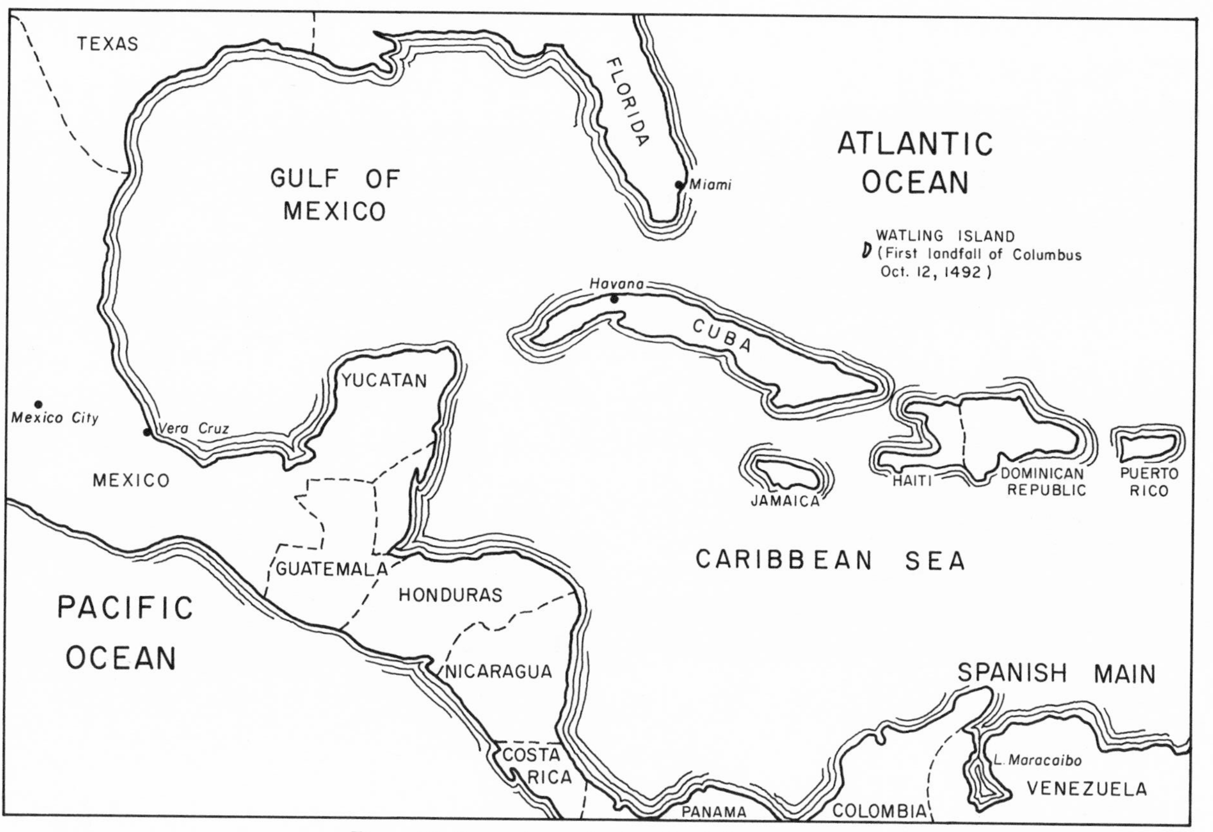

Fig. 10.2. Map of the Caribbean area.

lished a fort on Hispaniola and subdued the natives, shipping many prisoners in chains to Spain to be sold into slavery. These actions not only sparked the imagination of other adventurous souls to journey westward in search of treasure, they also inspired an era of plunder and savage brutality.

Columbus made four voyages to the New World. At the last attempt in 1502, he sought a passage through the islands to Asia, where the true Cipangu was believed to exist. A landing on the coast of Honduras revealed unexpected treasure. Here, the natives were well versed in metal mining and metalworking. They had developed methods of copper smelting and also of working veins of silver and gold occurring in the hinterland. After collecting much booty, the expedition moved southward and eventually broke up, with problems of ship maintenance and failing health and morale.

Meanwhile, other fortune seekers were setting sail to the New World. Some of these expeditions were mounted from Cuba, already settled as a Spanish possession. In 1517, Córdoba discovered Yucatán and a year later, Velásquez, who was the Cuban governor, dispatched his nephew there. He returned with a collection of treasure in the form of golden ornaments which were traded to him for trinkets by subjects of Montezuma, the powerful emperor of the Aztecs. Velásquez now sent Hernando Cortés to take the Aztecs and to colonize their land. After many friendly encounters with Montezuma, and much intrigue and desperate fighting, Cortés took Tenochtitlan (now Mexico City) and sent a tremendous hoard of gold to Spain. Unfortunately, there were two flaws in this approach. First, many of the priceless artistic golden ornaments were melted down into solid gold bars, worth about seven million dollars (but infinitely more as ornaments); and this treasure did not all reach Charles V of Spain. Some was captured at sea by French pirates and eventually found its way to Francis I of France.

The wealth, not only in raw gold and silver, but in delicately and exquisitely worked ornaments and other objects of art accumulated by the Aztecs and associated tribes, is difficult to describe. But it was seized and collected by the Conquistadores from all parts of the land. The Aztecs, who had dominated Mexico, were subjugated in 1521. Their land was taken over by the Spaniards, who rebuilt Mexico City. They developed the country and it became astonishingly rich in mineral wealth.

Mines in Mexico (New Spain) and in other parts of the New World were developed at an astonishing pace by the Spanish colonists, following the conquest. By 1524, tin and copper mines were opened up and smelting plants established in Taxco. Ten years later, silver mining was begun in Taxco, the first adit in the New World being officially declared open by Cortés himself. Actually, the first silver-mining operations were begun in 1525 at Morcillo, Jalisco, and at Villa de Espiritu Santo in Compostela, Nayarit (see figure 10.3).

The latter mine became the first Mexican bonanza in 1535. Mining at Zacatecas began in 1540, and during the next eight years important silver lodes were discovered at Guanajuato and Santa Barbara. Between 1552 and 1555, the famous silver veins of Pachuca, Fresnillo, Sombrerete, and Chalchihuites were discovered. The Real del Monte Mine was opened up in the Pachuca district of Hidalgo in 1551. In 1574, mining began in San Luis Potosí.

In 1554, in Pachuca, Bartolomé de Medina invented the patio process for extracting silver from its ores with the use of mercury. (This was later rediscovered at Virginia City, Nevada, in the 1860s and known as the Washoe process.) The five main stages in the process involved: (a) crushing and grinding the ore, (b) treading the wet pulp (mixed with salt, mercury, and roasted copper pyrite) by men or horses, (c) spreading it in large cakes on the floor of a patio, (d) washing the cakes with water in tubs with beaters to separate the silver amalgam, and (e) retorting the amalgam to recover the mercury. The whole cycle took up to four weeks. Several variations of this method were later developed, such as trampling in wooden troughs (at first cold, later heated), and forming large balls of the material, treated in an oven.

Just as Spain had formerly been the source of mineral treasure for the Phoenicians and others, now Mexico was to become the great treasury of Spain. Spanish prospectors had set out across the land and discovered rich gold, silver, lead, copper, and mercury deposits, such as those at Taxco and Guanajuato. The states of Zacatecas and Chihuahua became incredibly wealthy. But nearly every state in Mexico yielded astonishing mineral wealth, and Spain flourished accordingly.

Meanwhile, another Conquistador, Francisco Pizarro, had crossed the isthmus of Panama and had ventured southward along the west coast of the new continent (see figure 10.4). He later conquered Peru in 1533, defeating Atahualpa, the ruler of the Incas. Tremendous piles of gold and silver treasure were collected. Beautifully decorated goblets, jugs, trays, vases, utensils, tiles, and plates, and all manner of ornamental reproductions in gold and silver and gems were seized. Again, much of this was reduced to ingot form by melting, thereby destroying years of creative Incan craftsmanship. The value of this bullion alone was estimated to be worth up to 15 million dollars. One-fifth was set aside for the emperor; the generals and soldiery split the balance. Never before in history had military adventurers reaped such a fortune.

But this was not all. Pizarro now marched on the city of Cuzco, high in the Andes. Again the city was plundered and enormous spoils in golden treasure were divided among the

Fig. 10.3. Map of Mexico.

Fig. 10.4. Map of South America.

soldiery. But quarrels later developed. Pizarro himself was finally put to the sword by a band of conspirators in 1541.

The Peruvian Indians, the Incas, had developed a metalworking culture over a period of a thousand years. They had worked in bronze, using tin recovered in Bolivia. They were also masters in stonework: over 3,000 miles of roads and many bridges were built over deep canyons in the Andes.

Following the subjugation of Peru by Pizarro, important mines were developed in South America by the Spaniards. Mercury was discovered in Peru in 1558, and deposits were mined in Huamanga and Huancavelica. This obviated the need to import mercury from Spain. In 1571-72, Velasco adapted the mercury amalgamation process for the treatment of the silver ores of Upper Peru (Bolivia).

In 1542, gold deposits were opened up in Carebaya, La Paz, Oruro, and other places in Upper Peru, at an altitude exceeding 12,000 feet. Three years later, the first silver mine in Bolivia was discovered at Potosí, 3,000 feet higher, when a llama kicked away a bush to which it had been tethered, thereby exposing a rich vein of native silver.

Rich gold ore bodies were found in Confines and Quilacaja, Chile, in 1550. In 1584, King Philip II promulgated a summary of mining laws to regulate the mining industry in the New World. This did much to stabilize mining operations.

Mints for coinage production were established in 1535 in Mexico City, Lima (Peru), and Santa Fé (Colombia); also in 1572 in Potosi (Bolivia), and in 1606 in Zacatecas (Mexico).

Meanwhile, Europe was already booming with prosperity following the impetus given by the spirit of the Renaissance. However, the increasing demand for precious metals, not only for coinage but to serve the metal crafts industries, was being sorely felt. Fortuitously, the corresponding shortage was now able to be supplied from the New World. By 1550, Europe had received nearly 100 million dollars in gold and silver from this source, thereby doubling its existing supply of metal for currency purposes. This stock had risen to about 600 million dollars' worth by 1600, due to an additional flow from mines in Mexico, Peru, Chile, Bolivia, and Colombia.

Spain imported 200 tons of gold and 18,000 tons of silver from the New World between 1521 and 1660. Output from European mines consequently declined after 1550. In addition, extremely large amounts of wealth in gold and silver ornaments were coming in. Fortunes for individuals, mainly merchants and noblemen, were being created at a fast rate. All this had the effect of supporting the arts and developing commercial transactions.

In the long run, Spain did not gain her full share from this inflow of wealth. Most of it was hoarded, and so the boom produced an imbalance of trade. Spain soon lived beyond her means, importing too much from foreign countries, while her domestic production slowed down. Nevertheless, she had enjoyed a "Golden Age" in the development of art, literature, and science.

In 1557, during the reign of Philip II, Spain defaulted and became nationally bankrupt. In a desperate attempt to replenish the treasury, orders were issued to search for the old mines in Huelva province. Some of the greatest mines of the ancient world were rediscovered in the Rio Tinto area. A new policy was adopted to support domestic mining activities. Nevertheless, these efforts were frustrated by costly wars, corruption in high places, and a weak, confused government. The chief needs for the successful reopening of the remote Rio Tinto mines were adequate capital and a railway to the port. During the prevailing conditions, neither was available. Several adventurers during the next 300 years were granted mining concessions, but only two were even moderately successful for limited periods. They were severely circumscribed by government mismanagement and red tape. (At last, in 1869, the Spanish government retracted its long-held resolve for domestic operation of the mines and granted a freehold concession to the Rio Tinto Company, based in London, in 1873.)

Earlier Spanish imports had enriched France, England, and other European countries, whose economies were surging upward, supported, too, by the continuing effects of the Renaissance. Feudalism fell away still further, as more and more serfs were freed to ply their trades independently as artisans.

In his monumental treatise, Mining in the New World, Carlos Prieto underlines his thesis that mining has contributed enormously to world civilization. He comments on the incredibly short time during which the early exploration of the American hemisphere was conducted, and the impetus that mining gave to the development of agriculture and commercial activity, to the construction of roads, and to the building of cities and schools. He maintains that mining was the central development core of Latin America, and that the rapid exploration and development of the whole subcontinent in little more than half a century was due to the existence of gold. Some of his arguments are: (a) the search for gold and other precious metals had always been a developing force in literature, myth, and human life throughout the history of mankind; (b) though at the outset, there were excesses in the appropriation of treasures (plunder), we must keep in mind the attitudes of the period toward war and conquest; (c) discovery and exploitation of mines was no easy task; (d) the large shipments of gold and silver from the New World had a tremendous influence on the development of Europe; and (e) not all the precious metal was sent to the Old World. Much remained to form the basis for the development of Latin America.

Nevertheless, the discovery of extremely rich silver ores in Mexico, Peru, and Bolivia, following the advent of the Conquistadores, delivered a serious blow to silver mining in Europe after 1550. These developments indicated that gold and silver could be imported into Europe much less expensively than European miners could mine their own ores, now of much lower grade.

Of course, the early supplies of precious metals had been garnered in Mexico and Peru from streambeds. But large ore deposits were readily found and exploited by the Conquistadores, who used the natives as miners. In this way they made a tremendous contribution to the development of Latin American countries.

The slump in silver mining in Europe also brought about a reduction in the mining of copper ores. Mining activity in Europe did not revive for another two hundred years.

As a separate metal, zinc was unknown until the 16th century, although brass had been made much earlier by roasting copper ores with charcoal and calamine, a zinc carbonate mineral. Calamine had been mined by the Spaniards in Belgium since about 1400 but no records of mining were kept until about 1640. Calamine deposits also occurred in Algeria, Britain, France, Ireland, the Isle of Man, Sardinia, Silesia, and northwest Spain.

11 The Coal Age

The first evidence of the use of coal appears in the writings of Aristotle and Theophrastus. Later, coal was used by the Romans to a limited extent: a quantity of mined coal was found in one of the forts of Hadrian's Wall, in Britain (see figure 10.1).

In 1210, the monks of Newbattle Abbey near Preston, England, received a charter to mine coal. Coal was also being mined in Northumberland at about the same time. During the next hundred years, coal was mined in a number of areas in England and in Belgium.

Much of the coal mined in those days was from bell pits (see figure 11.1). Small shafts were sunk through the overburden to the coal seam, from which the coal was broken for some distance around the bottom of the shaft, and hoisted to the surface. As soon as roof falls or accumulations of water became intolerable, the shaft was abandoned and another sunk nearby. The resulting bell-like shape of the excavation gave a name to the method. Iron ore was also mined in this way. No doubt some coal was also mined from surface exposures (open pits) whenever possible.

The annual rate of production of coal in Britain until 1550 was no more than 200,000 tons. Up to this time, most industrial processes used wood charcoal for heating purposes, such as for lime burning, brickmaking, glass manufacture, brewing, and for the evaporation of seawater to produce salt. But by 1700 the denudation of nearby forests had brought about a shortage of firewood for charcoal production, and consequently coal was being substituted for charcoal. By 1700, Britain was producing 3 million tons annually.

A rapid expansion in coal usage occurred after 1700 because of the discovery that coal (and later, coke) could be used in the manufacture of iron. Both Sweden and Russia

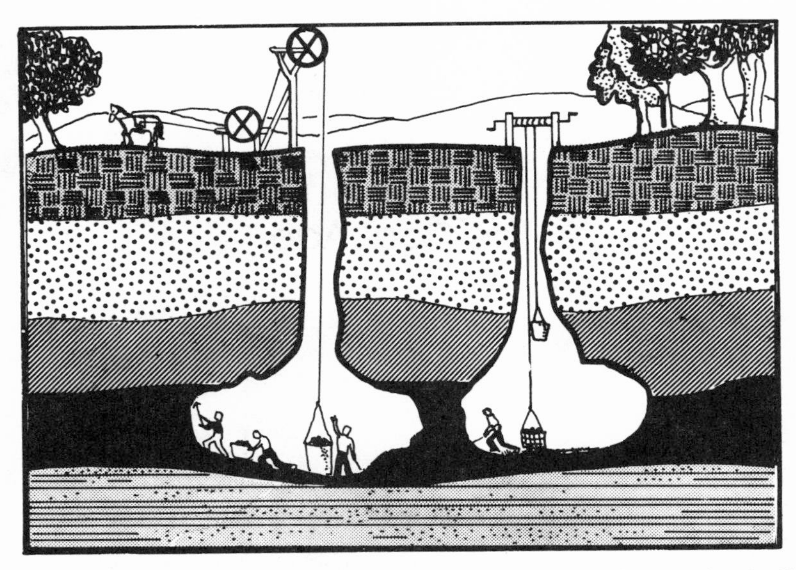

Fig. 11.1 Bell pits, showing whip hoist on left, and windlass (right).

began to use coal for iron smelting. In Britain, coal was first used to replace charcoal for smelting iron ores in the 17th century. Following the development of the use of coke (from coal) in the Derbyshire breweries, it was found in 1709 that coke could be used in a blast furnace to provide higher temperatures and therefore easier casting conditions. A better replacement for charcoal, coke was in general use for steel manufacture in 1745.

By 1760, there were 17 coke-fired blast furnaces in Britain. By 1775 this number had grown to 31, and to 81 in 1791. After this year there was a considerable growth in pig iron production in Britain, largely due to the greater blast pressures employed. The pig iron contained various impurities, mostly phosphorus.

The pace of coal mining was much slower in Europe. By 1700, as compared to an annual production of 3 million tons in Britain, only 400,000, 150,000, and 75,000 tons were being produced in Belgium, Germany, and France, respectively (see table 11.1 and figure 11.2).

Coke smelting of iron ore was introduced to France in 1785, Silesia in 1791, Belgium in 1823, Austria in 1828, and the Ruhr in 1850. As the use of coke spread in western Europe, both Sweden and Russia lost the lead they had established in the manufacture of iron, by smelting with coal.

Table 11.1. Coal Production in Various Countries (Includes anthracite, coal, and lignite - in round millions of tons)

Year	Great Britain	Belgium	France	Germany	United States	USSR
1550	**					
1660	2					
1700	3	**	**	**		
1750	5					
1800	10	4			**	
1820		6		1	**	
1850	56		5	6	8	
1854	65					
1860	80				20	
1866	103		12	29	29	
1871	117				40	
1895	193			104	175	
1900	229				270	
1910	268		36	118	502	
1913	292	23	43	278	563	
1925	243	23	65	272	592	24
1927				304		
1930					537	
1940					512	
1960	225				434	460
1970					613	
1972	132	8	36	510	602	722
1975			25	463	655	
1979	124	6	20		684	719

**Fewer than 1 million tons.

Note: China produced 24 million tons in 1925, 462 million in 1972, and 650 million in 1979.
Poland produced 35 million tons in 1926, 209 million in 1972, 212 million in 1975, and 234 million in 1978.

Further expansion in coal production after 1700 was somewhat restricted by political struggles in Europe and by the need to overcome technological problems in Britain as well as in continental Europe. These problems were centered around the need to improve methods for draining water from the mines, for hauling the broken coal along the galleries to the shaft, for hoisting the coal up the shaft, for providing ade-

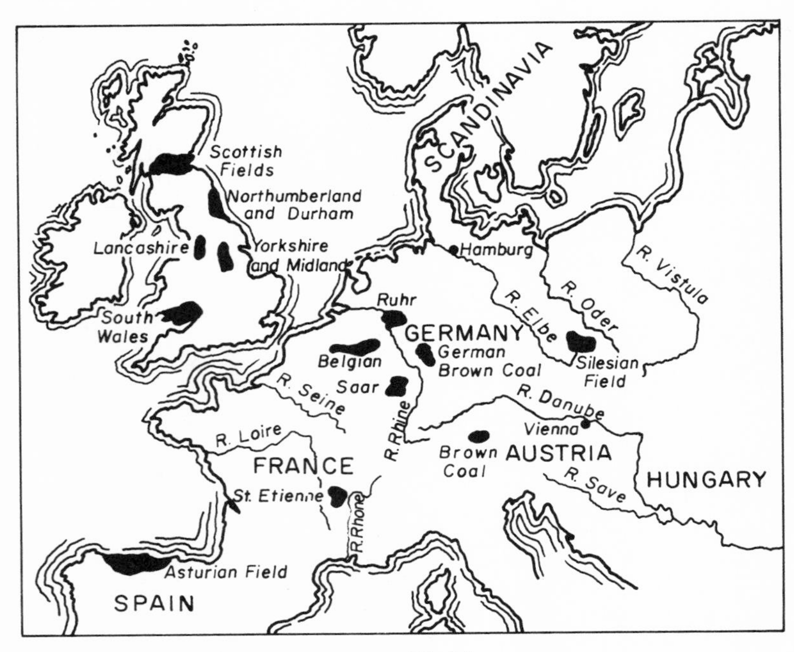

Fig. 11.2. Map of European coalfields.

quate ventilation, and for reducing the incidence of notorious disasters resulting from gas explosions. These aspects are dealt with in chapter 21.

Generally, the room-and-pillar method of mining coal was used at first (see figure 4.7). This involved the extraction of coal from rooms (bords, or stalls) advanced in the seam about 12 to 20 feet wide, leaving pillars 40 to 60 feet wide (and perhaps 50 to 100 feet long) in between the advancing rooms. The object of the pillars was to support the roof rock. The coal seam was, therefore, dissected into a series of rectangular blocks. In advancing each face, coal was extracted by first undercutting the coal manually. A horizontal kerf was cut into the coal near the floor across the face for a depth of about 2 feet by miners, or "hewers" using hand picks.

The upper face of coal was then induced to cave to this kerf by roof pressure, or by using explosives, first employed for this purpose in Britain in 1638. The resulting broken coal was then shovelled into transport units and hauled to the shaft.

Because only about one-third of the coal was extracted, the room-and-pillar method was regarded as wasteful. Efforts to extract the coal in the pillars at a later stage were not attempted or were not always successful. Nevertheless, the room-and-pillar method is still being used in some countries. Most of the current annual production of 300 million tons of coal mined underground in the United States today is won by this method, but of course with mechanized units.

In order to avoid the wastefulness of the room-and-pillar system, the longwall method was introduced into the Shropshire coalfields of Britain in 1770. This involved the mining of a long face of coal by a team of hewers. The working area near the face was protected by a line of wooden props supporting the roof. As the face advanced, the props were withdrawn and moved up progressively; the roof was allowed to cave behind the props. This area became filled with waste rock from the collapsed roof. The waste material is known as gob or goaf. With the longwall method, very little of the available coal tonnage is left behind unmined (see figure 4.7).

It is interesting to note that the longwall method was not introduced into the United States on an organized basis until about 1960; but its use is accelerating, in a mechanized version, copied from current European practice.

Britain led the way in coal-mining developments in the 18th and 19th centuries, just as Belgium was the leader in continental Europe. By 1830, Belgium produced six million tons annually from about 300 mines. The two main coalfields in France were much slower in developing. By 1850, France was producing somewhat less than five million tons per year.

Developments in Germany at this time were far more interesting. Her extensive coalfields were not fully exploited until 1850. The deposits in the Ruhr and the Saarland, as well as the Westphalian deposits near Aachen, were not effectively worked until 1815 and the Silesian field was not seriously developed until 1840. By 1820, German coal production was of the order of only one million tons (see table 11.1); and yet it had risen to six million tons by 1850. This rapid rise in output can be attributed to two main causes: the rapid development of railroads after 1835, and a series of favorable political decisions. By 1850, Germany had 23,500 miles of operating track, compared with only about 2,000 miles in France. The governments of the various German states had nationalized the railroads and the coal and iron mines, thereby fostering the development of these industries following the Industrial Revolution in Britain.

The Industrial Revolution took place in Britain in the 1770's following an interesting period of invention and research based upon the introduction of steam power for operating various production units that were worked previously by manual, animal, or water power. Furthermore, because of the exhaustion of firewood supplies, the use of coal for steam raising and for production of coke for steel-making had been an important factor.

The development of steam power was initiated by David Ramsay (1631), Thomas Savery (1699), and Thomas Newcomen (1712). All these inventions were based upon efforts to use steam power to remove the water accumulating in mines. Improvements were made later by other engineers; these developments will be covered in chapter 17.

The advent of the Industrial Revolution brought about a tremendous demand for coal for steam raising, and for coke in the smelting of iron and other base metals. Iron and base metals were required for making steam engines, other items of machinery, railroads, bridges, and associated appurtenances.

The benefits of the Industrial Revolution soon spread over Europe and across the world. If not for developments in steam engines for pumping water from mines, it is difficult to envisage how these industrial innovations could have occurred. Coal became the basis of modern industry.

The advent of the Industrial Revolution was most propitious for the United States, in that the Revolutionary War followed almost immediately. It provided a convenient springboard from which the United States was able to develop as a nation.

Coal was first used in America by the Hopi Indians in Arizona. They burned coal to fire their pottery. It was also discovered by French explorers in 1679 along the Illinois River.

Meanwhile, since the arrival of the first British immigrants in Jamestown, Virginia, in 1607, the American economy had been based mainly upon agriculture. Some early crude ironworks and forges had been set up along the eastern seaboard (following the first installation at Saugus, Massachusetts in 1646) using small deposits of local bog iron ores and powered by water wheels. Whereas most of these early furnaces and forges had been fueled with charcoal, the use of coal (first mined commercially near Richmond, Virginia, in 1745) later took over. Up to 1840, the United States had been smelting iron ore with anthracite (discovered in Pennsylvania in 1791) instead of charcoal; coke was first used after the Civil War (1865). Coke which is made from coal was at that time more readily available than charcoal which is made from firewood. As a result, coal production increased at an accelerated rate.

In the year 1800, there were 108,000 tons of coal produced. By 1850, production had grown to 8,356,000 tons, thereby surpassing the production of France and Germany. By 1866, Britain was producing over 100 million tons of coal per year (see table 11.1).

Meanwhile, iron, copper, tin, and lead were still being produced in Europe at a modest rate to serve the local industries. Most of these mines were in the Harz Mountains, the Ruhr, the Erzgebirge, Cornwall, and Falun, Sweden.

The various technological mining problems persisted in base metal mines, as they had done in coal mines, until the advent of steam power and Cornish pumps at the threshold of the Industrial Revolution. Apart from the need for these metals for agricultural and war purposes, the demand rose tremendously following the Industrial Revolution in the 1770s.

The introduction of the Cornish steam pump had been a great boon to the English copper-mining and tin-mining industry. By the 19th century, mining in Cornwall was at its peak. In Cornwall, the vein tin generally occurred below the copper deposits. Black tin output per year increased from 700 tons in Elizabeth's reign, to 2,000 tons in 1742, 3,000 tons from 1752 to 1778, and to about 14,000 tons in 1877. About 183 mines were being worked, 27 of them being alluvial operations. Cage hoisting had taken over from kibbles. German mining engineers from the Harz Mountains had introduced the use of explosives (1638) and the "man engine."

Output of other base metal mines increased steadily; but of course the economics of precious metal production in Europe was still overshadowed by the cheaper costs of mining high-grade deposits in Latin America. An example of this competition is given by the output of just one mine. The famous Cerro de Pasco Mine in Peru was accidentally found in 1630. It produced ore containing copper, silver, and gold. Within one year, the value of the silver alone reached 550 million dollars.

Further important developments in mineral production occurred in the New World during the Coal Age.

In Mexico, a great bonanza was discovered in the state of Hidalgo in 1762. About 31 million dollars worth of silver were extracted from a mine in the Pachuca district. Between 1748 and 1759 Don José de la Borda made extraordinary profits from his La Lajuela Silver Mine in Taxco. Famous silver mines were discovered at Catorce, San Luis Potosí in 1773. In 1792, the mining industry of Mexico had grown to the point where it was considered necessary to establish a mining school. Accordingly, the Royal Seminary of Mining was inaugurated in Mexico City with Don Fausto de Elhuyar, the discoverer of tungsten, as the first director. Two years later, he was succeeded by Don Andrés Manuel del Rio. He found vanadium in the dark lead ores of Limapan. From 1521 to 1922, the mines in Mexico produced 155,000 tons of silver valued at over three billion current dollars: well over 60 percent of the world's silver output.

In Bolivia, the mines of Cerro de Potosí produced silver to a value exceeding one billion pesos between 1545 and 1803. Don Alvaro Alonso Barba invented the "kettle and cooking process" for the amalgamation of silver-bearing ores in Tarabuco, Bolivia, in 1609. This process achieved amalgamation by heating to the boiling point the finely ground ore with water and mercury in copper kettles equipped with beaters for stirring. This reduced the time cycle of the cold patio process to a few days. In 1640, the first edition of his book, El Arte de los Metales, was published in Madrid. It was later issued 35 times in 6 languages.

In Brazil, gold had been known to exist in the province of Minas Gerais as early as 1543. But it was not until 1693 that large lodes of ferruginous quartz gold ore were exploited and the famous mining town of Villa Rica de Ouro Prêto was established. A mint was set up here in 1725. Other gold-mining operations were commenced in 1719 at Mato Grosso and in 1724 at Araguaia. By 1726, diamonds were discovered at Tejuco (now Diamantina). An English company in 1830 formed the St. John del Rey Company to operate the famous gold mine at Morro Velho. The value of the vast output of gold and diamonds mined in Brazil during the 18th century can only be guessed at because records were kept unreliably. From 1700 to 1800 it is estimated that a thousand tons of gold were produced, most of it between 1740 and 1760.

Tin in veins had been discovered in 1710 on Banka Island and later at Billiton, Indonesia. In 1793, alluvial tin deposits were being worked by Chinese operators.

In about 1580 A.D., during the political attacks from the Emperor Akbar, the Maharana of Mewer (Rajasthan, India) had to seek refuge in the hills. He and his family were saved by being lowered to safety in wicker baskets into the tin mines of Zawar.

In north Africa, salt had been mined in Taodeni, Mali, since 1585 A.D., in an area about 400 miles north of Timbuctoo. Similarly, gold was produced from the Wangara Mine and transported via Bamako and Timbuctoo to Fez.

Only in Siberia was gold mining able to compete with Mexico and South America before 1850. In fact, the Russian goldfields yielded 60 percent of the world's gold until 1848 (see figure 11.3). In 1744, a quartz outcrop was dicovered near Ekaterinburg in the Ural Mountains. Gold has been mined from this deposit ever since. Placer deposits were also found in the Urals in 1771 but were not systematically exploited until 1814. Since then, gold production from these sources has expanded significantly.

Placer deposits in the Yenisei Valley, Siberia, were discovered in 1838, leading to a gold rush. By 1851, about 20,000 miners were working there in 106 gold mines. In 1846, the largest gold strike was made near Bodaybo, on the river Vitim, a tributary of the Lena (see figure 11.3).

A tremendous upsurge in all aspects of metal mining was now about to take place. During the second half of the 19th century, new mines were to be opened up in other continents.

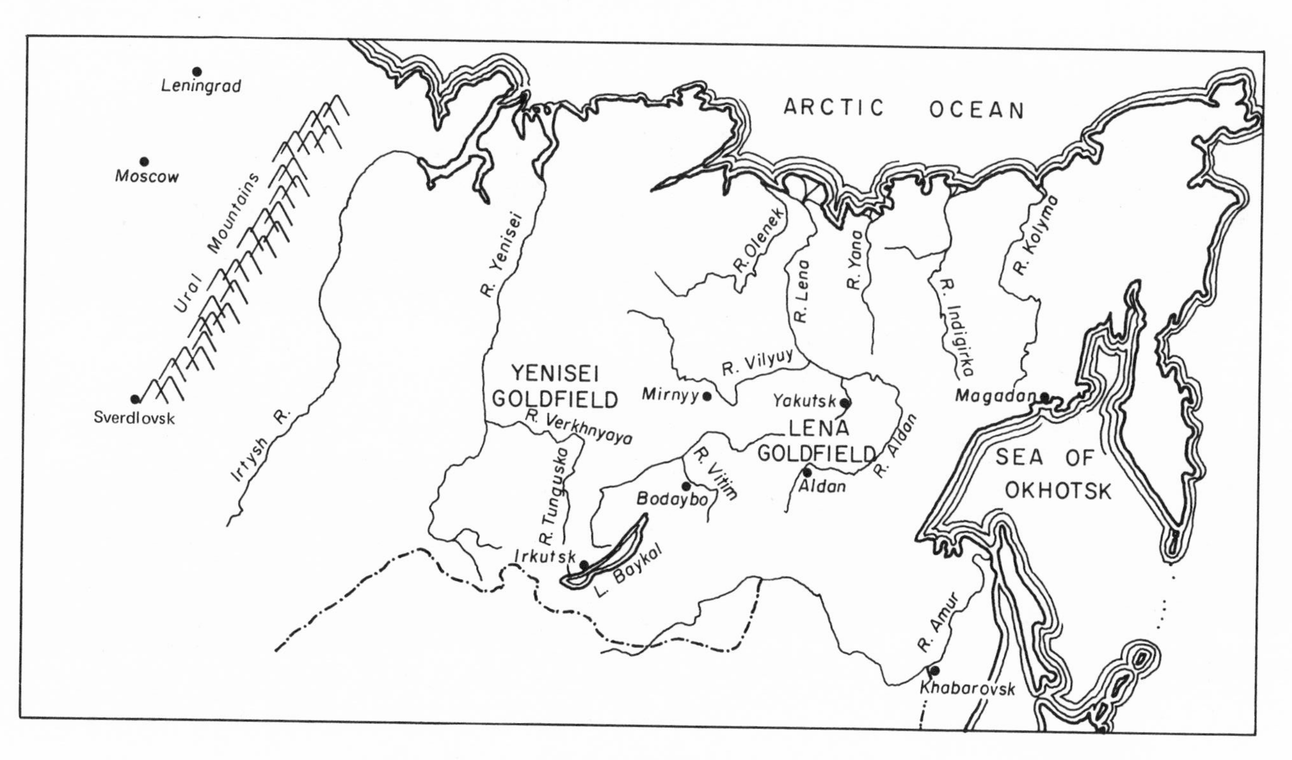

Fig. 11.3. Map of Russia and Siberia.

12 The Petroleum Age (from 1850)

PETROLEUM

Although various hydrocarbon compounds such as crude bitumen and asphaltum had been discovered and used by the ancients as far back as the 9th century B.C., and seepages of petroleum and asphalt were known to the South American Indians, and the pitch lake on the Island of Trinidad was noted by Sir Walter Raleigh in 1595, the discoveries of oil deposits by drilling were not made until 1857 in eastern Canada and Romania and until 1859 by Edwin L. Drake near Titusville, Pennsylvania.

By 1900, mineral oil had been found also in Russia and Indonesia. In 1908, British geologists discovered a large oilfield in Iran. The first Arabian oil well was spudded in by 1935. Until the development of the automobile, in about 1900, this mineral oil was used mainly in the form of kerosene for oil lamps, or as lubricants. At an earlier stage animal oils had been used for these purposes.

Since then, petroleum (and natural gas) have become extremely important sources of energy. Petroleum derivatives are more convenient to use than coal, especially in mobile vehicles. Today, petroleum is one of the most important items in world trade and in international transport.

Apart from its important uses as a fuel, it is also, like coal, a basic raw material for the production of a wide range of chemical products. These include plastics, lubricants, and synthetic rubber (petrochemicals).

Since 1900, the world demand for petroleum and natural gas has grown at an exponential rate. The search for new sources, known as oilfields or oil reservoirs, proceeds at a fast pace in most countries, in sedimentary basins both on land and on the continental shelf. The Soviet Union and the United States possess the most extensive sedimentary basins but the largest known petroleum reserves have been developed in the area around the Persian Gulf.

Over the years since 1900, oil production has come mainly from the United States, the Middle East, Venezuela, Indonesia, and Russia. More recent discoveries have been made in Libya, the North Sea, Nigeria, Angola, Colombia, Australia, Gabon, Chile, Bolivia, Peru, Romania, China, and Mexico. It will, therefore, be seen that petroleum exists in a wide area of the world, although many countries produce no oil at all.

For several reasons, oil exploration and production is now being carried out offshore in many areas, in depths of water of 600 feet and more. Most of the world's production comes from wells 1000 to 15,000 feet in depth, and exploratory wells are being drilled as deep as 30,000 feet.

Apart from the occurrence of oil in the liquid and gaseous phases in stratigraphic traps in the earth's crust, large deposits of rocks known as oil shales and tar sands have been found in the United States, the Soviet Union, Canada, China, Sweden, and Australia. These latter deposits might eventually be explored further, after existing oilfields have become depleted.

Natural gas occurs both associated with and separately from petroleum oil. It can now be traded internationally as liquefied natural gas (LNG) in special ocean-going tankers.

Changes in the production sites and development of reserves of petroleum are taking place rapidly across the world. New deposits are being discovered month by month, just as older deposits are becoming exhausted.

COAL

Three important developments in coal mining occurred during the Petroleum Age. These were: (a) the emergence of America and the Soviet Union as the leading coal-producing countries, (b) the mechanization of coal production methods, and (c) the challenge of new fuels, such as petroleum and natural gas.

After 1895, Great Britain ceased to be the leading coal-producing country (see table 11.1). Production in America almost doubled in each decade until 1910; then the rate of increase slackened following the expansion of oil refineries to produce gasoline and other refinery products.

The most revolutionary change in mining methods was the replacement of manual undercutting with the pick by the introduction of coal-cutting machinery. Apart from this, open pit (strip) mining, first employed in the European lignite fields, was used increasingly to mine bituminous coal. There had been little scope for this in Great Britain where the coal seams seldom occurred near the surface.

The early pioneering work in the development of coal-cutting machinery had been undertaken by British engineers. As early as 1761, Willie Brown's Iron Man was used, a mechanical robot designed to follow the actions of a hewer but with stronger and more frequent blows of the pick. Later, a similar principle was used but the effort was provided by a horse. Improved methods involved the use of a circular steel disc, acting like a rotating knife blade, to cut the coal. The main problem in all these devices was the need for a suitable form of power to actuate them.

Nevertheless, by 1850, Lord Cochrane had introduced the use of compressed air to replace steam which had many disadvantages when used in underground situations. The first coal-cutting machine operated by compressed air was introduced in 1861. It was a circular disc set horizontally, with a number of hardened pick points mounted on the periphery. The disc diameter varied in different models from three to six feet. The first successful cutter was used in the Lancashire coalfield in 1868. In the United States, where the coal seams were more regular and thicker, compressed air operated cutting machines were in use after 1877.

In the 1860s, a bar-type machine was also introduced. It consisted of a circular tapered bar set with teeth or cutters. It was more applicable to soft friable seams where it was disadvantageous to cut a continous kerf. For these conditions, a disc-type undercutter was impracticable. After 1900, electricity was introduced to replace compressed air for the operation of coal-cutting machines; but its use was not readily accepted at first because miners were suspicious of its safety, and mine owners were not eager to face the additional capital outlay.

In 1913, only 8 percent of coal was mechanically cut in underground coal mines in Britain. This figure had risen to 14 percent by 1921 and 31 percent in 1930. Britain had many small producing units and capital was not readily available for these small owners for purchase of machinery. Apart from this, the thin irregular seams in Britain were not so readily applicable to the use of coal-cutting machinery.

From the bar-type cutter, American engineers developed a successful chain-type machine by 1894. By this time the use of electricity was more general, and American coal mines became mechanized at a fast rate. By 1920, 60 percent of underground coal was mechanically cut. By 1929, this figure had risen to 78 percent.

Since 1900, and the introduction of electricity, there was a general move in Europe and America to mechanize coal-breaking operations at the face. The regular procedure was a cut-drill-blast-load cyclical operation. In the longwall method, props were reset, coal was cut and drilled, and shotfirers blasted the coal at the end of the day shift. On the afternoon shift, coal was loaded out. On the night (graveyard) shift, pack walls were built, stable ends were advanced, the conveyor was moved up to the face, and the roof was generally kept maintained.

Therefore, coal was produced on one shift per day only. With the pressure of an ever-reducing margin between costs and revenue, an incentive was provided to improve this cyclical system. The aim was to produce coal on all three shifts, in an system referred to as "continuous mining".

The first step was to use a cutter-loader type of machine, which cut and loaded the coal onto the conveyor belt in one action. This enabled coal to be produced on two shifts. The third shift moved the machine forward and secured the roof. Finally, a machine was introduced to cut the coal and load it onto the face conveyor, so that both the cutter and conveyor advanced as the faces were mined. For this system, a new type of roof support was developed. This was the hydraulic prop. It was a telescopic jack that gave rigid support to the roof. By operating a valve, it could be unloaded and set up in a new position. Props were therefore retracted and reset to the roof as fast as the face advanced. These early hydraulic props were later developed into integrated self-advancing systems.

These and other technical advances made cyclic mining obsolete. Every shift became productive. By 1950, all the main producing longwall mines in Britain and Europe were equipped with these powered support systems.

Nevertheless, not all coal seams could be worked in this way. Where seams were thin or discontinuous or inaccessible, such as in Britain and continental Europe (including some parts of Russia), coal had to be broken by pneumatic picks and hand-shovelled onto the conveyor belt.

Meanwhile, American mines had been using the room-and-pillar method. The same argument held: there was a need to abandon the cyclical system and to establish a continuous system of coal production on all three shifts. In order to achieve this aim with the room-and-pillar method, it was necessary to abandon the conventional method of cutting, drilling, and blasting, as had been done with the longwall method in Europe. This was not achieved until after World War II, however, when the use of tungsten carbide alloy hard-metal pick points became available. This hard metal had been invented and introduced by the Germans during the war for machining gun barrels.

The use of hard-metal pick points was the basis for the development of the "continuous miners", a highly integrated mobile machine provided with maneuverable cutting heads and a central loader-conveyor system. Different styles of directional cutting heads were pioneered and produced by several American manufacturers. The coal, broken from the face, fell onto a loading apron and was conveyed to the rear of the machine for discharge onto a transport vehicle. Such a machine was introduced by 1950. This new system avoided the need for the cut-drill-blast-load cycle. In fact, no explosives were involved.

Developments in surface mining of coal in the United States were spectacular during this period. The introduction of open pit mining in Illinois in about 1886 began with plowing (ripping) the surface, scraping away the overburden, and exposing the coal for removal manually by pick and shovel. Machinery was introduced by 1910 and steam shovels with a dipper capacity up to eight tons were in use. The stiff overburden was first loosened by drilling and blasting.

The first large-scale effort to mine coal by surface mining had been made in Germany in the lignite mines. This is where the giant bucket-wheel excavators were developed for the purpose (see figure 4.11). Overburden waste was dumped into worked out areas. Lignite was worked on a single long face up to 2 miles long and up to 300 feet high, depending upon the thickness of the seam. Most of the seams were 30 to 60 feet thick. Another machine was the bucket-chain excavator, which ran along a bench on rails, continuously clawing the coal up the face and discharging it at the top onto a conveyor belt. Both the bucket-wheel excavator and the bucket-chain excavator are now in general use for lignite mining throughout the world, in capacities up to 260,000 cubic yards per day.

ALUMINUM

Not until 1845 was aluminum metal produced in the laboratory. At this time, it was regarded as an exotic metal available only at precious metal prices. It was derived from high-grade bauxite ore and digested with caustic soda to produce alumina. This stage was later improved by K. J. Bayer. The next stage, to reduce alumina to the metal, was invented independently in 1886 by C. M. Hall in America and by P. L. T. Héroult in France. This involved the present-day method of electric smelting. Cheap electrical energy was necessary. This practice was first initiated in Switzerland at the Rhine Falls at Schaffhausen. Later smelters were installed in Norway, in the state of Washington (U.S.A.), in Canada, and in Tasmania, based upon the availability of hydroelectric power.

OTHER METALS AND INDUSTRIAL MINERALS

Petroleum, coal, and aluminum were certainly not the only minerals or metals being developed during the Petroleum Age. Discoveries of the older precious and base metals occurred across a wide area of the world, following a tremendous surge of exploration activity. In order to classify these discoveries, it is better to consider them on a country-by-country or regional basis.

America

At the beginning of the Petroleum Age (1850), five main mineral discoveries launched the United States into an era of industralization that later developed her status as the most powerful industrial nation of the 20th century. These discoveries were: gold in California in 1848; high-grade iron ore in the Marquette Range, Michigan, in 1844; high-grade copper ores in northern Michigan in 1841; major lead deposits in the Mississippi Valley, from 1864; and petroleum, from 1859.

All of these discoveries (and that of coal, found earlier) were to result in significant output of the raw materials necessary as a base for the huge industries that were later to develop. (In 1929, America produced 35.87 percent of the world's coal, 67.58 percent of the oil, 47.72 percent of the steel, 48.78 percent of the copper, 35.56 percent of the lead, and 39.28 percent of the zinc: in short, nearly one-half of the world's basic materials).

Precious metals

It was Californian gold that created the impetus and sparked the great movement westward to develop the rich land area that was to become part of the United States. The other four discoveries necessarily took some years to materialize and were attended with much less public fervor.

Figure 12.1 shows where, at Coloma, on the bank of the south fork of the American River, about 40 miles above its junction with the Sacramento, James W. Marshall, with a crew of workmen, was building a water-wheel driven sawmill for Captain John A. Sutter, in 1848. On the morning of January 24 he noticed shiny flakes of metal in the water passing through the tailrace. This was later found to be gold. Sutter tried to keep the discovery quiet, at least until the crew completed its work. But the news leaked out and several of the crew deserted to dig for gold. By May 1848 the whole territory of California was swept with gold fever.

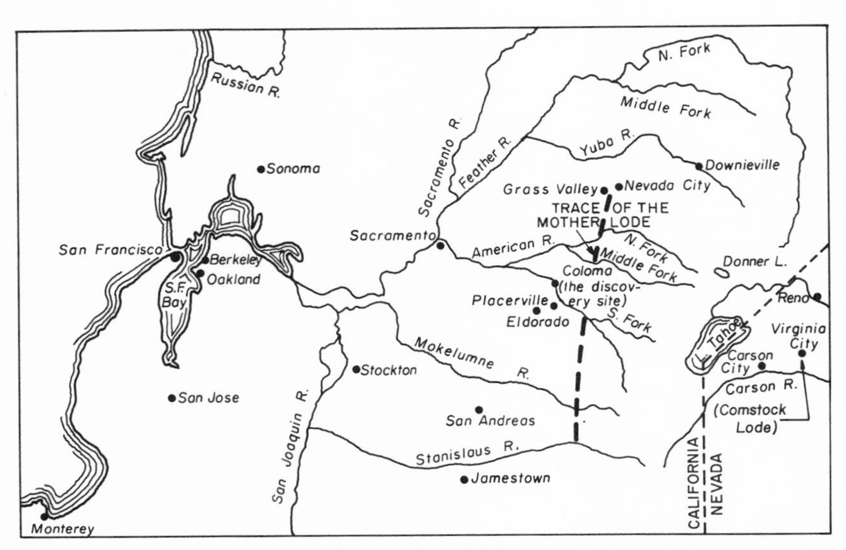

Fig. 12.1. Map of California goldfields.

The California territory was to be ceded from Mexico under the Treaty of Guadalupe Hidalgo only nine days later than the date of the actual discovery. Neither of the signatories knew about the gold discovery. America invested 15 million dollars in the treaty; but the value of gold won in the next 11 years amounted to 595 million dollars.

The news of spectacular finds and the hope of gaining sudden wealth spread like wildfire across the country. When the news reached the eastern seaboard, people downed tools (and pens) and commenced the great rush westward. They knew that little expertise was needed to win gold from these placer deposits; and few tools were required beyond a pick, a shovel, a gold pan, and a few camp utensils. It was a heaven-sent opportunity for the greenhorn to amass a fortune, with few assets beyond the tools, the will to work and to withstand crude living conditions, a strong heart, and a soul full of hope. There was plenty of the latter, the main element that characterizes a gold rush, or motivates any prospector, or for that matter, any mining investor.

But the first bridge to cross was a difficult one: how to get there?

There were three main "bridges" leading to California from the eastern seaboard, and all were long ones, wrought with hazard. The survival rate was low. But this did not deter the oncoming swarm of treasure seekers.

Some took overcrowded sailing ships for a journey of 1,800 miles around Cape Horn. Others went by sail to Panama and crossed the isthmus on foot or mule along the cobble stoned trail through the jungle, beset by insects producing tropical fevers, which caused many deaths. (This was the trail - El Camino Real - built by Pizarro to transfer the golden treasures of Peru to the waiting Spanish galleons before sailing along the Spanish Main to Spain. It was likewise used by Henry Morgan, the English pirate who sacked the old city of Panama. Morgan missed the huge golden altar that was surreptitiously painted black in good time by an enterprising priest. This famous altar was later transferred to a small unimposing church in the modern Panama City, where it may now be seen). The survivors then waited for an opportunity to catch another ship to San Francisco.

Then there were those intrepid adventurers who took the long overland "bridge" along rough trails across the plains and mountains from towns now known as Kansas City and St. Joseph. Most followed the Oregon Trail as far as Soda Springs, Idaho, and then branched along the California Trail. Others split off to the southwest through Utah along the old Spanish Trail. Thousands of wagons (prairie schooners) headed westward along and across difficult terrain. In some places they had to be ferried across rivers on makeshift rafts or lowered by ropes down steep cliffs. Thousands of people died from cholera, Indian attack, or just plain sickness or accident. The hardships endured were beyond description. Travel could take place only between April and September: 123 days to bridge a gap of about 2,100 miles.

By 1850, the population of the goldfields had grown to 100,000. Enough gold had been won in two years to pay for the Treaty of Hidalgo three times over; and California was admitted as a state of the Union by September.

Most of these early treasure seekers had little or no knowledge of mining, but this was not necessary. Gravel shovelled from the bedrock under existing or former streams carried particles or nuggets of gold. Some miners developed crude sluice boxes, rockers, or cradles to separate the gold from the lighter gravels. The Long Tom was later introduced. It was a more sophisticated style of long sluice box that could be fed with gravel by a large team of workers. By 1873, hydraulic sluicing with high pressure nozzles was well advanced. About 775 water races, totalling 4,863 miles in length, had been constructed in 25 counties.

But it was not only the "easterners" who rushed to the field to try their luck. By 1853, about 45,000 Americans, 15,000 Mexicans, 25,000 Frenchmen, nearly 20,000 Chinese, and about 5,000 other foreigners were on the field. Life was rough in those days. Hard physical labor and simple diets in crude shelters in all weathers were the rule. Illnesses such as diarrhea, dysentery, and scurvy were common. Murders were frequent. The crude drinking saloons were a source of leisure entertainment. Those who were lucky provided song; others were able to drink to their sorrows.

Following the first flush of feverish activity, by which time most of the streams had been scoured, those who were experienced sought the source of the placer gold. The mother lode was found first at Mariposa, then at Nevada City and Grass Valley in 1850 as a quartz outcrop, streaked with native gold. Later this lode was found to extend for 120 miles southward along the foothills of the Sierra Nevada where it fed the tributaries of the Sacramento and San Joaquin rivers (see figure 12.1).

At Grass Valley, gold quartz ore was discovered in June 1850. The Empire Star group of mines produced over 120 million dollars' worth of gold over a period of 100 years from 200 miles of underground workings developed from the Empire Shaft. (This shaft had reached an inclined depth of 11,007 feet by 1956. In 1957, there was a labor strike and the mine closed and was sold at auction in 1959).

The gold ore, as mined, was crushed at first in a Mexican arrastra, later in Chilean mills, and still later in steam-driven stamp mills. The gold was collected by amalgamation, the mercury having fortuitously been produced from ore mined elsewhere in California. By vaporizing and distilling the mercury in cast-iron retorts, the gold was recovered from the amalgam. But only about one-quarter of the vein gold was free milling: amenable to amalgamation. So the chlorination process was introduced in 1860 and cyanidation in 1918.

Further producing mines were established along the southward extension of the mother lode. Some of these were: the Bunker Hill, the Keystone, the Central-Eureka, the Lincoln (or Union) from which Leland Stanford built his financial empire, the Argonaut, the Kennedy, the Utica, and the Carson Hill (the richest of them all, producing 26 million dollars' worth of gold). These mines were located at various settlements south of Grass Valley, such as Coloma, Plymouth, Sutter Creek, Angel's Camp, and Carson Hill.

Captain Sutter died penniless, as did James Marshall, at the age of 74. At the discovery area, at Coloma, in a state park and museum site, there is a tall bronze statue of Marshall, pointing to the historic spot which has meant so much to America.

The gold rush to California was not necessarily the last effort of the unlucky ones. Gold fever had sparked an interest in other areas of the west, such as in Colorado, Nevada, Idaho, Montana, British Columbia, Washington, Dakota, and Arizona.

In 1859, gold and silver were found in an area near Carson City, Nevada, when a ground squirrel displaced the thin cover from a rich vein. At first, the find was not attractive because while gold was being washed in pans and in sluice boxes, a certain amount of "blue stuff" clogged up and clouded the washing process. This discouraged the prospectors. It was later identified as argentite, a rich silver mineral, and this new knowledge set off a rush. The deposit became famous; it was known as the Comstock Lode in a town to be called Virginia City.

About 35 mines were established on this lode over a length of four miles. The great Bonanza Mine, as its name implies, was the most productive. The lode varied from 100 to 200 feet wide and dipped generally at an angle of 45 degrees to the east. One-third of the precious metal was gold, and two-thirds silver, with some antimony present.

By 1861, there were 10,000 miners and camp followers. In 1863, these mines produced silver and gold to the value of 12 million dollars. By 1872, the yield had increased to 135 million dollars. In all, over 400 million dollars of wealth was produced. But most of the rich bonanzas occurred deep underground in the lode. These would not have been discovered but for the capital invested by financiers to provide plant to work the deeper deposits and to develop the metallurgical techniques for extracting the gold and silver. Several of these financiers made fortunes. Among them were John Mackay, in whose name the Mackay School of Mines at the University of Nevada was founded and George Hearst, whose name is well represented at the University of California campus at Berkeley.

George Hearst had had experience as a hard-working miner in Missouri before he came to Grass Valley to mine quartz veins. He later became an original stockholder in the extremely rich Ophir Mine. One of the finest halls on campus is the Hearst Memorial Mining Building, erected by his widow, Phoebe Apperson Hearst, who financed it in his honor. In the main foyer is a bronze-plaque bearing the bearded likeness of George Hearst (1820-1861) and the following inscription:

> This building stands as a memorial to George Hearst a plain honest man and good miner. The stature and mould of his life bespoke the pioneers who gave their strength to riskful search in the hard places of the earth. He had warm heart towards his fellow man and his hand was ready to kindly deed. Taking his wealth from the hills he filched from no man's store and lessened no man's opportunity.

Phoebe Hearst also underwrote a competition for the design of an enduring architectural plan for the campus. In turn, her son, William Randolph Hearst, built a gymnasium in 1927 as a memorial to his mother.

The Comstock Lode was an extremely wet mining area. Stupendous quantities of water had to be pumped out to keep the mines free of seepage water. To overcome this problem, the Sutro Tunnel, a drainage adit over four miles long, was driven into the mountain below the ore bodies to drain the mines to a depth of 2,200 feet.

In the Ophir Mine, it was difficult to support the walls of the excavation in the huge stopes. A system of modular timber sets was designed by a young German mining engineer, Philip Deidesheimer, to overcome this problem. This system of "square sets" is now adopted as standard practice around the world whenever such conditions are encountered.

Techniques developed in other parts of the world were eagerly adopted by the mining companies at Virginia City. Some of these were the patio process, the diamond drill, the compressed air rockdrill, dynamite (in 1868), and steel wire hoisting rope. These advanced techniques were then available to other mining centers developed later in the western states.

The Comstock gold rush had similar features to those in other areas. Those who arrived early with great hopes were either well rewarded, or as in most cases, bitterly disappointed. Mining was carried out under rough conditions. The early depletion of alluvial gold was followed by the development of underground vein mining in hard rock, calling for crushing of the ore and metallurgical extraction of the precious metals. Opportunities for the single individual to become wealthy were then necessarily transferred to the capitalists.

Two outstanding personalities figured in the California and Comstock rushes. Mark Twain, a reporter for a local newspaper, did not profit from his investments in the mines at Virginia City. Lola Montez (1818-61) apparently did so, in a physical sense. She was an Irish girl who was proficient in Spanish dancing, and became a paramour to Franz Liszt and to King Ludwig of Bavaria, from whom she acquired the title "Countess of Landsfeld". Later, she started a career as a high-class harlot at the former Indian summer capital at Simla. The French dubbed her "la grande horizontale." In 1853, she transferred her activities to San Francisco where her dancing career was shortlived. She moved to Grass Valley, and in 1855 to the Ballarat mines in Australia. She was a very colorful figure, having the reputation of dazzling "two lovers, two husbands, and the King of Bavaria." Lola Montez died penniless in New York in 1861. Her old home, now restored as a historical landmark, may still be seen on Mill Street in Grass Valley.

Meanwhile, the feverish interest in searching for gold and silver did not abate. Prospectors and others spread out across the west, panning the various streams and testing the outcrops (see figure 12.2).

Fig. 12.2. Map of western United States.

In Colorado, a rich strike was made in 1859 by George A. Jackson near Blackhawk, where a smelter was established in 1868. This was followed soon after by John S. Gregory in the gulch that became Central City, a mining town that has been continuously preserved. Central City is now a famous tourist mecca; apart from the old Opera House that still stages performers of world renown each season, such exhibits as "The Face on the Bar-room Floor" are monuments of interest to tourists. Another famous silver mining town in the vicinity was Georgetown, where rich silver veins were found in 1860.

In the 1860s, gold and silver were discovered higher up in the Rockies at Leadville. Later, lead carbonate ore was found; it was extremely rich in silver. There was a rush here in 1878. The Matchless Mine turned storekeeper H. A. W. Tabor into a millionaire nine times over. Many interesting sagas have been written about his exploits. (Leadville is now the base of American Metal Climax for mining molybdenite ore). It was also at Leadville where Meyer Guggenheim first came to sell shoe-laces, and later developed his luck into a huge investment empire, mainly by establishing custom smelters around the west to smelt the various parcels of ore sent to them by prospectors and small operators. Copper and zinc, too, were produced at Leadville.

Across the range from Leadville, a series of fabulous new finds were made at Silverton and at Telluride, where the world's first alternating current power line was put into service. At Telluride, as the name implies, most of the gold was found to occur, not as free native gold, but as various tellurides of gold.

In 1860, Pierce found gold near the south fork of the Clearwater River in Idaho. Other finds were made at Florence in 1862. Several gold discoveries after 1862 were made in Montana. In the following 12 years, placer gold to the value of 150 million dollars was won from several areas, first at Helena, the state capital since 1875, then in Virginia City and Bannack. At Butte, "the richest hill on earth" was discovered. It was first regarded as a gold strike (a placer deposit), in Silver Bow Creek, in 1864. But this was mined out three years later. Then the black siliceous outcrops on the hill were prospected and the gold rush became a silver boom, still later a copper boom. Hundreds of claims were staked, but eventually these were all taken over by the Anaconda Company, mainly as a copper mine (see later).

As mentioned earlier, the Comstock Lode was the first prolific producer of silver in the United States; in many of the later discoveries of metalliferous ores, rich oxidized silver minerals were found in abundance at or near the surface outcrops. One such example was the Silver King Mine found in 1871, in the Superstition Mountains of Arizona. A few years later, silver was also found in Tombstone, in which area 80 million dollars in silver were produced within a decade or so.

One of the outstanding gold strikes of the late 19th century occurred in 1876. A rich find of placer gold was discovered in the Black Hills of South Dakota by Moses Manuel. About 25,000 excited miners rushed into Deadwood Gulch in an area of Indian land where Wild Bill Hickok and Calamity Jane made frontier history. This led to the opening of the Homestake Mine in 1877. This famous mine is still working, now more than a mile below the surface, and producing over 20 million dollars' worth of gold per year.

The value of gold production in the western states in 1877 was recorded as: Arizona, $350,000; California, $15,000,000; Colorado, $3,500,000; Montana, $3,550,000; and Nevada, $18,000,000.

In 1881, a bonanza was found in the melting snow near Juneau, Alaska, This led to the development of the famous Alaska-Juneau Mine, which produced over 34 million dollars in gold, and in 1882, to the Alaska Treadwell Mine which produced gold to the value of 64 million dollars. In 1917, these great underground mines were abandoned when sea-water broke into the workings.

In the 1890s, prospectors streamed into Cripple Creek, Colorado. By 1894, this goldfield became a hive of activity. In 1895, a treatment plant was erected here. Cripple Creek was the richest goldfield in the United States by 1900. Within a radius of ten miles, more than 500 underground mines existed. At one stage, 440 were in operation at the same time. Three railroad companies extended their tracks into the town, and a train either arrived or departed every 6 minutes on an average throughout the 24 hours. Here, too, the gold was in the telluride form. Many millionaires were created by the wealth won from Cripple Creek.

The last great gold rush of the 19th century occurred in 1898, just over the Alaskan border into the Canadian Yukon Territory. It is said that two woodsmen (Robert Henderson and George W. Carmack) spotted gold in the Klondike River while fishing it for salmon. The news broke out in 1897 and a rush developed, similar to that of California in 1849. The chief mining center was established at Dawson City, which within a year had a population of 25,000. By early 1898 a fleet of 41 ships was plying between San Francisco and Skagway, the nearest port of disembarkation to the goldfields. From here, the prospectors had to cross the ranges by either of two passes, the Chilkoot or the White Horse. The Canadian Northwest Mounted Police checked 22,000 people through the frontier during the winter of 1897-98. There was a continuous stream of prospectors climbing the 4-mile long approach to the pass, each carrying his own food supply. Beyond the pass, a 500-mile journey over lakes and rivers, following the thaw, was required to reach Dawson City. Thousands died along the way. Those who survived found it worthwhile. They first concentrated on surface gold. But the gold-bearing gravel was frozen under a thick coating of moss. The miners had to adapt themselves to the harsh, wintry conditions. They developed a shallow series of underground drifts and freed the gravels by fire-setting.

In the first three years, the area yielded about 2.5 million dollars' worth of gold. But by 1900, the more easily won gold had been recovered. It was now time for plant and machinery to replace the efforts of the individual.

Two other areas brought enormous riches in silver (and gold) to the state of Nevada. At Tonopah, a boom occurred and within a space of less than 30 years, following its discovery in 1900 by Jim Butler's burro which accidentally discovered the Mizpah vein, the underground mine produced silver and gold to a value of 153 million dollars.

In 1902, further to the south, in the Nevada desert, some famous gold veins were staked in the area known as Goldfield. From the Mohawk Mine, a single 47-ton carload of ore netted $574,958 after deducting freight and smelter costs. The Goldfield district contributed over 90 million dollars to the wealth of the country.

The California gold rush of 1848-49 created the impetus for the industrialization of the United States. The American west was within a few short years combed by prospectors and settlers; and the development of the West was aided by the feverish extension of several railroad systems to serve the new settlements and to profit by the opportunities thereby created. The great plains states were opened up for agricultural development, and the mining of iron ore, copper, and lead (all discovered a few short years earlier in the Lake Superior and the Mississippi Valley region) was soon about to reach a steady production rate and reinforce the march to industrialization that the gold rush had sparked.

But not only in the United States was the impetus of the California gold rush felt. So great was the degree of hope kindled in the minds of prospectors that a tremendous surge of exploration activity was engendered in many other countries. The second half of the 19th century set the stage for a vast array of fabulous producing mines around the world, many of which are still operating, as we shall see later.

Copper

The earliest and most significant copper deposit had been reported by the state geologist Douglass Houghton in 1841 in the Keeweenaw Peninsula of Upper Michigan (see figure 12.3). It had originally been worked by an ancient race of Indians in about 4000 B.C. They had also mined salt, flint, and turquoise from caves and outcrops. Alexander Henry had started mining in the Ontonagon area in 1771 but he was not very successful.

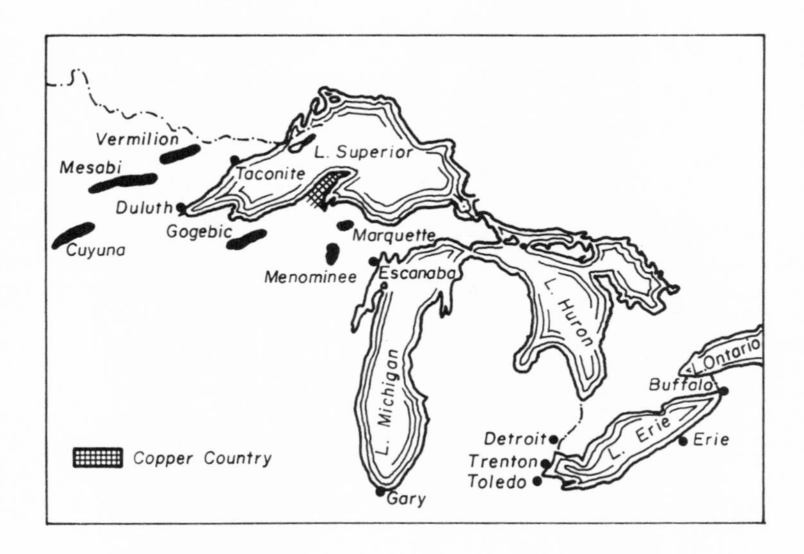

Fig. 12.3 Map of Lake Superior copper-mining and iron-mining areas.

Copper was found to occur in two types of deep-seated formations in this area. In 1856, the Quincy Mining Company sank several shafts along the 45 degrees dip of the amygdaloid lode. By 1930, one of these had reached an inclined depth of 9,300 feet, about 6,000 feet vertically below the surface. The famous mammoth Quincy steam hoist, with bi-cylindro-conical drums 30 feet in diameter is still to be seen here, as a museum piece. Many other mines were also set up on this lode.

In 1864, the Calumet Mining Company opened up a mine in a conglomerate formation. It became the Calumet and Hecla Consolidated Copper Company in 1923. By 1930, these consolidated mines had produced copper to the value of over one billion dollars and had distributed 185 million dollars in dividends to their shareholders. The original promoters and the board of directors were mainly Bostonians. Most of the miners were Cornishmen, Swedes, Finns, and Irishmen.

By 1930, the total contribution of the copper country to the national wealth was 4 million tons of copper. Shareholders had been paid a total of over 340 million dollars in dividends from the amygdaloid and conglomerate mines. Tremendous nuggets of native copper were found in this area. One weighed over 400 tons. The copper country played a large part in laying the foundation of America's industrialization.

Most of the mines ceased operation in the 1930s, but there still remains a possibility of reopening some of them at enhanced copper prices. Further to the west, the White Pine Copper Mine is presently producing a significant amount of copper.

Extensive deposits of copper ore were found later in other parts of the United States. As mentioned earlier, the silver veins at Butte, Montana, had been traced to a depth of 300 feet by 1878, and here they became rich in copper content. In 1877, one thousand tons of ore were produced, averaging 35 percent copper and 15 ounces per ton of silver. In fact, Butte has been dubbed "the richest hill on earth." It has produced three billion dollars of mineral wealth and is still in operation (both on surface and underground).

Another famous ore deposit was the United Verde at Jerome, Arizona. Between 1887 and 1913, a total of nearly 500 million dollars' worth of copper, gold, and silver were extracted from the two main mines. Other copper mines have been producing for many years in Tennessee, Utah, Nevada, New Mexico, and Arizona.

The Bingham Canyon Mine in Utah was originally mined for gold, silver, and lead, but in 1896 its potentiality for copper production was recognized. Yet the copper mineralization, though extensive, was thought to be too sparsely disseminated to be payable. A mining engineer, Daniel C. Jackling, reported 12,385,000 tons of 2 percent proven copper ore and 25 million tons probable. At the time, this was too low in grade to warrant extraction. However, Jackling pressed his views that such a low-grade ore could be made payable by using large-scale mass-production open cut methods. The Utah Copper Company was formed in 1903 and during the next ten years produced 67 million tons of 1.4 percent copper ore. By 1930 the ore reserves had increased to 640 million tons and the company had mined copper, gold, and silver ores to the value of nearly 850 million dollars; it had paid 200 million dollars in dividends and 60 million dollars in taxes.

This mine is now owned by Kennecott Copper Corporation; it has produced two billion dollars' worth of metals. In 1974, this unit mined 35,277,300 tons of ore, of an average grade of less than 0.75 percent copper, and 106,290,805 tons of waste overburden - over 386,000 total tons per calendar day. In 1977, the ore reserves were quoted as 1.6 billion tons of 0.7 percent copper ore: enough to give a further life of 40 years. About 600 billion tons of rock (ore and waste) have already been mined. The original hill has been mined out and the open cut is regarded as the largest man-made hole in the world. This famous open cut is now 2.5 miles long, 1.5 miles wide, and over half a mile deep, with more than 60 working benches. It is a stupendous attraction for tourists.

Other notable copper mines in the United States, worked by the open cut method, are the Chino, Tyrone, and Santa Rita (started in 1804 by a Spaniard) in New Mexico; the Morenci, Silver Bell, Twin Buttes, Pima, Bagdad, Sierrita, and Ray in Arizona; and the Yerington, Ely, and Weed Heights in Nevada. Underground copper mines have included the Miami, the Globe, the Ruth, the Inspiration, the Magma (discovered in 1875), the San Manuel, the White Pine, and those at Ducktown, Tennessee, where black copper oxide was discovered in 1849. Many of the above mines are in full operation. The famous Copper Queen Mine at Bisbee, Arizona, was staked in 1877. It produced about 12.5 billion dollars' worth of copper, lead, silver, and gold before it closed in 1975.

Iron ores

Until 1856, steel was made by the carburization of wrought iron. As new processes were developed by Bessemer and Siemens to produce better quality steel at lower prices, the rising demand for iron ore continued. The early bog iron ore deposits were soon discarded. Magnetitic iron ore was worked in Pennsylvania in 1732. But it was in 1844 that the Marquette Range deposit of high-grade iron ore was discovered. From 1852, supplies of iron ore were sufficient to justify the establishment of a major steel industry. In subsequent years, other immense deposits were discovered and placed in production. These were Menominee (1872), Vermilion (1877), the Gogebic (1884), and the Mesabi Range (1891) deposits. The ore occurred in thick horizontal layers no more than 200 feet below the surface.

These huge deposits offered scope for mechanized production to strip off the overburden and expose the ore for extraction, largely by steam shovels. Large companies with great financial resources were formed to develop the mines and the steel industry.

The United States Steel Corporation owned and mined 65 percent of the Lake Superior ores; it also owned five large docks and an extensive fleet of lake steamers, operated its own railroads, and produced half of the American steel output. These iron ore deposits of the Lake Superior region have produced on a prolific scale: more than 70 million tons per year of ore averaging over 60 percent iron. This represents from 80 to 90 percent of the country's production.

The iron country, along with the copper country, the coal mines, and the immense wealth in gold from California and the West, triggered the industrial development of the United States. The consequent development of railroads also opened up the country for agricultural and commercial development.

America made 4 million tons of pig iron and steel in 1880 as against 8 million in Britain and 2.7 million in Germany. By 1895, America was the world's largest producer.

Nevertheless, in the course of time the high-grade iron (and copper) ores of the Lake Superior region became exhausted. By 1950, much lower grade iron ore was being mined, and alternative deposits were feverishly being sought. Two solutions to this problem were found.

1. After 1948, the low-grade (30 percent Fe) iron ore was beneficiated (upgraded to 60 percent) and produced in a suitable physical (pelletized) form for feeding to blast furnaces; and
2. enormous deposits of iron ore were discovered in Canada along the Quebec-Labrador border, centered on Schefferville (see figure 12.4). These ores were shipped by a new railroad system to a port on the St. Lawrence River and then by lake steamers to the great steel cities of Erie, Cleveland, Toledo, and Gary.

Meanwhile, a certain amount of iron ore was mined from the Pilot Knob and Iron Mountain deposits in Missouri.

Lead

Widespread interest in searching for other minerals in the United States followed the finding of extensive deposits of gold, copper, and iron ores in the middle of the 19th century. The most important of these was lead; it should be noted that silver minerals are usually associated with occurrence of gold, lead, and copper. In addition, zinc minerals are also closely associated with those of lead.

The early explorers and hunters depended upon supplies of lead for their ammunition, Lead ore was mined and smelted in Virginia in 1621, and lead mining commenced in a small way along the Mississippi River in 1690 near Dubuque, Iowa.

Then in 1720, lead ore was mined in Missouri. By 1828, lead mining commenced near Platteville, Wisconsin. In the 1860s tremendous reserves of lead ore were developed near Bonne Terre, Missouri and the St. Joseph Lead Company was formed.

At about the same time, lead (and zinc) ores were found in southwestern Missouri. By 1888, the Eagle-Picher Mining Company was being formed to operate mines near Joplin. Extensions of this lead-zinc belt were later found in neighboring Kansas and Oklahoma. The whole area became known as the Tri-State Mining District. As much as 24 million tons were being mined annually at the beginning of the 20th century.

Lead, as mentioned earlier, was mined, along with silver, in 1880 at Leadville, Colorado; relatively small, rich discoveries were made in other districts, as in the Park City and Tintic districts of Utah. But the major discovery of lead (and

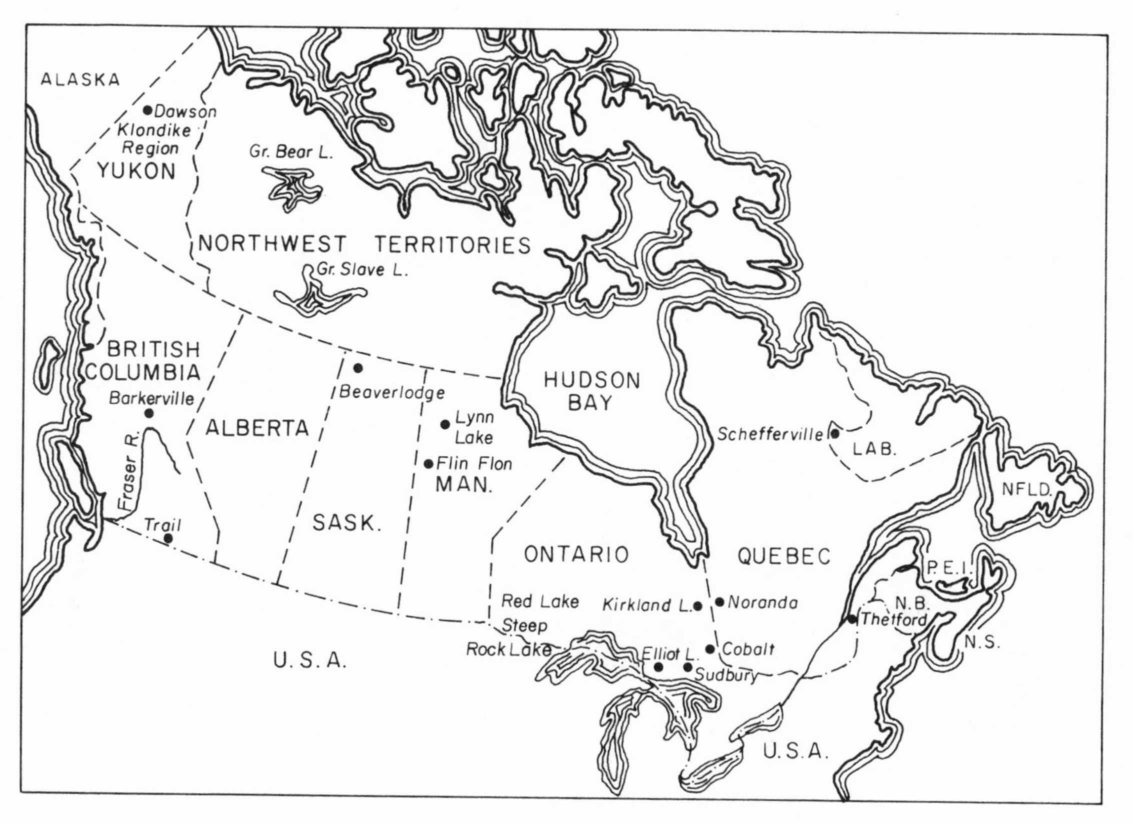

Fig. 12.4. Map of Canada.

silver) was made in the Coeur d'Alene Valley of northern Idaho. In 1883, rich silver-lead ore was first disclosed by the wayward kicking of a burro near Kellogg. This fabulous district has produced tremendous wealth; it is still producing today from several mines down to 7,500 feet below the surface. All of these mines produce lead, silver, and some zinc, but a few produce exceptionally large outputs of silver, such as the Sunshine and the Lucky Friday Mines.

Other Metals

The world's largest deposit of molybdenite ore was developed in 1917 (and is still in operation) by the American Metal Climax Company near Leadville, Colorado. Zinc mining began in 1860 in Illinois and Pennsylvania, and later in New Jersey, Wisconsin, Tennessee, Missouri, and Colorado. Otherwise, much of the zinc produced is won as a co-product of lead mining. A notable deposit of ores of tungsten and other rare metals was opened up at Pine Creek, near Bishop, California.

Nonmetals

Some significant discoveries of nonmetallic mineral deposits may also be listed. These include the discovery of phosphate deposits in South Carolina (1867), in Florida (1888), in Idaho (1906), and in North Carolina (1950); potash deposits in New Mexico (1930s) and in Utah (1950s); sulphur and rock salt deposits in the Gulf Coast salt domes; the extensive deposits of borates in Death Valley, California; the trona deposits of Wyoming; and rock salt in New York and Michigan.

Near Murfreesboro, Arkansas, diamonds were discovered in an old volcanic pipe in 1906, in what is now known as the Crater of Diamonds State Park. The diamond field covers an area of 78 acres in which over 60,000 stones have been found, the largest weighing 40.23 carats. This is the only known diamond field in North America. It is not being commercially operated.

Canada

Although coal was discovered in Nova Scotia in 1720, mineral prospecting in Canada did not reach a significant level until 1842 when the Geological Survey of Canada was established.

A wild gold rush occurred at Barkerville (in 1862) in the Cariboo district of British Columbia, following the Fraser River discoveries in 1858. The gold vein found by Barker produced 600,000 dollars' worth of gold. In its heyday, Barkerville was the largest town in North America north of San Francisco and west of Chicago (see figure 12.4). The whole Cariboo district produced 45 million dollars in gold. Another rush occurred on the Chaudiere River in Quebec. On the Kootenay River in southeastern British Columbia the world-famous Sullivan Mine was opened in 1894, yielding lead, zinc, silver, and other metals. A smelter was established at Trail in 1896 to serve this famous mine.

Apart from these discoveries, other mineral deposits were found before 1900. These included the copper and lead finds in Newfoundland, the great asbestos deposits around Thetford, Quebec, in 1876, and a number of copper, silver, and iron deposits in Ontario; and of course the Klondike gold rush, mentioned earlier, occurred in 1897. Mining activities again boomed after 1900, especially following the discovery of spectacularly rich gold-silver-cobalt-nickel deposits at Cobalt, Ontario, in 1904 (see figure 12.4).

While blasting for a railroad cutting in northern Ontario, the great nickel-copper deposits of the Sudbury basin were found in 1884, and put into operation in 1910. In 1902, an efficient process for the separation of nickel and copper, and a new use for nickel, in the production of nickel-steel alloys, had been developed by a Scottish scientist. From this time onward, the mining of nickel ores received an encouraging impetus. The Falconbridge Mine was discovered in 1918 and opened in 1929. These developments were followed by the finding of huge gold deposits in the Porcupine area, such as at Timmins and Kirkland Lake (1910-13). The production of such mines as the Hollinger, the McIntyre, the Dome, and the Wright-Hargreaves made significant contributions to Canada's development as a mineral-producing country. Major copper-gold discoveries were then made at Noranda, Quebec, followed by gold deposits at Red Lake in Ontario in 1919.

During the 1920's a steel industry was established, and various metallurgical innovations were made in the treatment of complex ores, including the refining of zinc and nickel. This enabled the important copper-zinc deposits at Flin Flon (discovered in 1915) and at Lynn Lake and Snow Lake, Manitoba, to be developed.

Gold discoveries were made at the Giant Yellowknife Mine, on the shore of the Great Slave Lake in the Northwest Territories in 1935, and later, and other large base metal deposits were also found and developed in this area by several operators. Uranium deposits were found and worked in 1930 at the Great Bear Lake and at Blind River, Ontario. The Steep Rock Lake iron ore deposits were developed in 1943 by draining the lake. In the 1950s, tremendous tonnages of iron ore were developed along the Quebec-Labrador border.

By 1975, Canada ranked first in the free world output of asbestos, potash, nickel, and zinc; second in gold, silver, and molybdenum; and third in copper and lead. In fact, about 90 percent of the world's nickel was being produced in Canada.

Australia

The Californian gold rush and its effect on the national industrial development of America was by no means an isolated global event. Although small pockets of gold had been found in New South Wales in 1835 and 1841, one of the unsuccessful prospectors in California (Edward Hargraves) returned to his native Australia and found gold near Bathurst in 1851 (see figure 4.15). The discovery of rich placer gold at Clunes and at Ballarat, in the state of Victoria, happened a few months later. A tremendous rush was now centered upon Ballarat. Immigrants soon arrived from many parts of the world, including many who were unlucky in California. Enormous problems were experienced in maintaining law and order.

Methods of gold recovery devised in California were now used in Australia, but the scarcity of water in Ballarat precluded the use of the Long Tom. Instead, a puddling tub, made from half a barrel, was used as a rocker to break down the stiff clay from the alluvial deposits. The tub was half filled with pay dirt and some water from the creek, and the contents were then "puddled" with a spade until the clay lumps were disintegrated. As each tub became heavily charged with clay "soup," it was decanted and a fresh supply of pay dirt and water added. In time, all that remained was clean gravel and sand, along with the flakes and nuggets of gold. Many of the alluvial deposits at Ballarat were buried, being known as "deep leads." These were mined by shallow underground methods through shafts 120 to 160 feet deep, and usually under wet conditions.

In 1851, Australia had about 438,000 people. By 1861, there were about 1.2 million. During the 1850s almost 40 percent of the world's gold output came from Australia. In this area in 1869, the world's largest gold nugget, the "Welcome Stranger," was found. This nugget weighed 2,332 troy ounces, valued in those days at $40,000. Many of the mineral museums of the world display a replica of the "Welcome Stranger" nugget.

By 1853, deep underground mining activities had been centered in Bendigo where later there were 53 shafts deeper than 2,000 feet and one more than 5,000 feet deep. By 1872 there were 1,200 small mining companies operating at Bendigo. Mining of deep leads at Ballarat was by now virtually completed.

In 1854 there was an armed uprising by Ballarat miners who rebelled at the government restrictions on mining. As a result, the Victorian government revised its gold-mining laws in 1855. On payment of one pound sterling per year for a document called a "miner's right," a miner was entitled to dig for gold, to vote at parliamentary elections, and to have a voice in making his own mining laws.

In 1876, the total gold production in Victoria amounted to 400,000 ounces of alluvial gold, and 600,000 ounces from reef mining. About 1,100 quartz reefs were being worked. There were about 42,000 miners including 11,000 Chinese. The largest output was in 1879 when 2,985,991 ounces were recovered.

But, of course, the individual miner who could make a personal fortune in placer deposits was unable to compete with mining companies who had command of the capital financing that was so necessary to develop and equip deep underground mining operations in hard rock. Some of the mining companies, however, let "tributes." These were a form of contract made with a group of Cornish miners (tributers) to work a section of the mine on payment of a tribute, representing a percentage (10 to 30) of the gross value of the gold recovered.

Many of the individual prospectors moved into New South Wales to the district where Hargraves had found gold in 1851 near Bathurst. In 1872, spectacular gold finds were made in the Hill End district. But even this discovery did not put an end to the peregrinations of the individual prospectors. Placer and vein gold were found and mines were worked along the eastern coastal strip of Queensland. Mining operations yielded substantial wealth in gold in such areas as Kiandra (discovered in 1860), Clermont (1861), Gympie (1867), Ravenswood (1868), Charters Towers (1872), Palmer River (1873), Mount Morgan (1883), and Croydon (1885). The Palmer River goldfield was discovered by J.V. Mulligan. Since then, no great alluvial deposits have been found in Australia.

Charters Towers was the most productive goldfield in 1891-96, mainly because it was able to attract the necessary capital from Britain to sustain its activity. It reached its peak in 1899. Mount Morgan mined gold of over four ounces per ton from the oxidized zone in 1889. By 1901 it had become a copper-gold ore body.

Five suction dredges were in operation in the Beechworth alluvials of Victoria by 1891. In the same year the cyanide process was first used in Australia, at Ravenswood, north Queensland. It was later to be improved at Kalgoorlie. The Tasmania Gold Mine at Beaconsfield produced about 850,000 ounces of fine gold between 1877 and 1914, when it ceased operations due to an excessive water inflow. Australia was the largest gold producing country in the world in 1903.

Alluvial and vein tin mining were carried out in the New England Tableland district of New South Wales and southern Queensland, at Tingha, Tenterfield, and Stanthorpe. At the same time, extensive prospecting activity was taking place in Tasmania. The world's largest lode tin mine at Mount Bischoff was discovered in 1871, and tin is still mined from that same general area, at Renison Bell.

By 1886, gold was discovered at Mount Lyell, but it has since become a great copper mine. Pyritic smelting, developed by Lawrence Austin and Robert C. Sticht (a graduate of Clausthal School of Mines) in Montana between 1887 and 1896, was introduced by the latter at Mount Lyell in 1896. J.S. MacArthur, the Scottish co-inventor of the cyanidation process, was a director of one of the mines. In the 1890s, rich deposits of silver and lead were discovered at Zeehan. Lead is still being mined from a deposit at Rosebery, Tasmania.

In the meantime, one of the most fabulous silver-lead-zinc deposits of all time was found at Broken Hill in 1883 in the far west of New South Wales, following copper finds at Cobar (1870), about 350 miles west of Sydney. By 1891, the Broken Hill Pty. Co. had paid two million dollars in dividends. The Broken Hill field is still a prolific producer. The three main companies today operate at a high level of efficiency, and have contributed greatly to the social development of the city of Broken Hill (The Silver City), placed as it is in a desolate area.

It was in Broken Hill in 1904 that the famous flotation process was invented, although the Germans had been working on it since 1877, for the concentration of graphite. It was patented in the United States by the Minerals Separation Co. in 1905, but it was not successfully used there until 1911 when a plant at Butte, Montana was installed. By this time, about eight million tons of sulphide ores had already been floated. It has since been employed all over the world for concentrating many ores (mostly sulphide minerals). With the use of differential flotation, developed in 1922, the lead sulphide (galena) particles were also easily separated from the zinc sulphide (sphalerite) particles in the finely ground ore, thereby enabling two clean concentrates to be produced. The galena concentrates are shipped by rail for 150 miles to the famous BHAS Smelters, originally established in 1912 at Port Pirie, where refined lead ingots are produced, along with refined silver and gold bullion as by-products. Following a reorganization plan in 1925, this smelting plant became the largest in the world. Dr. G. K. Williams introduced the world's first continuous refining process for the separation of lead from zinc and other metals in 1932. The sphalerite concentrates from Broken Hill are also railed to Port Pirie, and then shipped as calcines to Risdon in southern Tasmania. Here, a modern electrolytic refinery, established in 1921, produces refined zinc slabs (some to a purity of 99.995 percent of zinc), along with cobalt and cadmium as by-products.

The wealth produced from the Broken Hill mines has intensified the industrial development of Australia in many ways. The mining companies themselves have, by both vertical and horizontal integration, invested in all manner of enterprises in primary and secondary industry to support the development of the Australian nation.

Yet the city of Broken Hill is established in a desolate, uninviting area of sparse vegetation, high ambient temperatures, and poor rainfall. Even though two large reservoirs were built before 1900, they were often completely dry. The city is now supplied with water through a 70-mile pipeline from a dam near the Darling River.

Broken Hill, in its early days, was well-known for its dreadful dust storms. In the 1930s, an enterprising worker, Albert Morris, convinced the Zinc Corporation Ltd. that they should fence off a barren sandy area on their southern boundary. He believed that the natural vegetation had no chance of developing because it was heavily grazed by local livestock. The mere fencing of this area soon turned a desert waste into lush natural pasture, which has further been developed by the company into a leafy park with pleasant gardens, lawns, and even fruit trees. Broken Hill is now relatively free of dust storms, as a result. This area, once shunned by man, is now the site of many of the homes of the mining staff. This noble experiment has since been repeated by other mining companies, and by the city council. Yet all these activities were instituted and completed 30 years before the spirited campaigns of the environmentalists of the United States in the 1960s.

The Ballarat gold rush of 1851 played a similar part in the triggering of industrial development throughout Australia, just as the California finds had done three years before in America. And just as the discoveries of copper (and iron and lead) in the Mississippi Valley and Lake Superior region were "waiting in the wings" to support this impetus, we find a parallel situation in Australia.

Apart from the discovery of coal near Newcastle in 1797 and small veins of silver-lead ore near Adelaide in 1841, rich copper deposits had been opened up in South Australia at Kapunda (1844) and also at Burra Burra in 1845. Copper was produced at Burra Burra in great quantitites from an open cut and underground workings, to a depth of 600 feet. However, this mine ceased operations in 1877. Meanwhile, new copper discoveries had been made at Moonta and Wallaroo in about 1860. All these mines, at Burra, Moonta, and Wallaroo, were staffed by immigrants from Cornwall. These "Cousin Jacks" made a tremendous contribution to the development of Australia. In fact, Moonta has often been described as "Australia's Little Cornwall." These mines closed down in 1922.

Widespread copper deposits were also discovered in 1867 in the Cloncurry district of northwest Queensland by Ernest

Henry. Despite many years of vicissitudes, including the unwise use of capital funds in some instances, this field did not respond to the hopes of the original investors. A narrow gauge railway reached Cloncurry from the Pacific coast port of Townsville, via Charters Towers, in 1908.

In the meantime, prospecting was continuing all over this vast, dry continent. Following the gold discoveries in north Queensland, prospectors surged westward and made further finds at Hall's Creek (1886); at various places in the Kimberley Ranges and in the Pilbara district; at Nannine, Meekathara, and Cue in the Murchison district; and in 1888 at Southern Cross. In 1892, Bayley and Ford discovered gold near Coolgardie, in the West Australian desert. This was followed a year later by another discovery at Kalgoorlie, 23 miles to the east, by Paddy Hannan. A large gold rush developed. All kinds of people downed their tools and surged by all available means of transport to the new goldfields. Some are credited with pushing a wheelbarrow, containing their possessions, for up to 1,000 miles across the desert. Kalgoorlie became the flourishing capital of the goldfields. Claims were staked for miles along the line of the lode. The central portion was known as the "Golden Mile." The fabulous Great Boulder Mine was staked and floated in 1893. Until 1963 this mine produced gold to the value of 100 hundred million pounds sterling, at 1970 prices.

A great deal of capital expenditure was involved, but the mines yielded fabulous profits. Some are still being worked today, to depths approaching 4,000 feet. Kalgoorlie had one particular problem, however: the availability of fresh water. The area is studded with dry salt lakes. Early entrepreneurs set up stills, not to produce bourbon, but fresh water, which sold at a higher price per gallon than whisky. Hundreds of these wood-fired condensers provided limited amounts of high-cost water, and yielded fortunes to the operators.

It was obvious that gold milling plants could not be set up under these conditions. But there was help at hand. The premier of Western Australia was a far-sighted statesman. A special reservoir was constructed by the government in the Darling Ranges near Perth (the capital), and water was to be pumped through a pipeline 351 miles long to Kalgoorlie. This was in 1898. Sections of the pipeline were 30 inches in diameter; and others were 36 inches. Seven steam pumping stations were established along the route. Pipes were made from steel plate rolled into a cylinder. The longitudinal seam was made with a special locking-bar joint, because welding processes were not then developed. A great storm of public criticism arose because of the taxpayers' money spent on this experimental "white elephant," which actually was the first large-diameter pipeline to be installed anywhere in the world for any purpose. Before water flowed through the line, the chief engineer commiteed suicide, because of the welter of public criticism. But the pipeline was a tremendous success. It was officially opened in a ceremony in 1903, and it has since made possible the development of the goldfields. Many branch lines along the route have been extended to serve farming and other mining communities. The pipeline is presently being duplicated to meet the increasing demand for water.

Before and after the discovery of Kalgoorlie, additional gold strikes were made in areas to the north, south, and west. In 1931, the famous Wiluna Mine, almost in the center of Western Australia, was placed in production, but it closed in the 1940s. The West Australian goldfields had contributed much wealth to the state, and to the nation.

It was not until 1908 that iron ore and steel production appeared. A deposit of iron ore was opened up in 1901 in the Middleback Ranges of South Australia to supply a small amount of flux for the Port Pirie lead smelters, established on the eastern shore of Spencer Gulf. This deposit soon exhibited a great tonnage of high-grade (65 percent Fe) ore, sufficient to form the basis of a steel industry. Within the next few years, the Broken Hill Pty. Co. (the first of the silver miners at Broken Hill) erected blast furnaces and a complete steel mill at Newcastle, New South Wales, where opportunities for developing port facilities existed and where the base of the coal industry had been situated since 1800. This plant was in production by 1915. Coke ovens were also set up. Iron ore was shipped by sea from this deposit, known as Iron Knob, to Newcastle. Later deposits in the vicinity were known as Iron Monarch, Iron King, and Iron Prince. For many years, these deposits fed the steel mills; and the industrial development of Australia accelerated.

But mining activity slackened off by 1914, and following World War I it was subdued until the revaluation of gold during the Great Depression caused a gold boom. Old gold mines were reopened everywhere, particularly in Western Australia. Rich gold ore was also found at Tennant Creek in the 1940s in the Northern Territory. One of the mines in this area, the Peko Mine, developed at depth into a significant copper producer. The old copper mines at Mount Morgan and at Cobar were reopened in the 1930s, largely because the copper ore carried a significant amount of gold.

During the 1920s, a low-grade lead-zinc ore body was being developed west of Cloncurry in Queensland, at a place now called Mount Isa. A mill and smelter were also built. Although the price of lead dropped to its lowest level ever in 1931, the mine commenced and continued in production during seven years of operating losses. This company paid its first dividend in 1947. Mount Isa Mines Ltd. is now the largest single underground mine in Australia and produces large quantities of lead bullion, zinc concentrates, and blister copper. Modern equipment, facilities, and methods compare with any in the world. It currently mines about 20,000 tons of ore per day.

Other mining developments during this period include the large scheelite deposit on King Island, found in 1904, and the Aberfoyle tin and wolfram mine, opened in 1926, in Tasmania.

South Africa

Until the early 17th century, South Africa was virtually uninhabited except for small numbers of nomadic aborigines (Bushmen), who lived in a state of savagery. Following the trail of early navigators such as Bartholomeu Diaz (1486) and Vasco da Gama (1497), and after winning their religious independence from Spain, both Dutch and French settlers arrived at the Cape of Good Hope and began to occupy areas of hinterland for farming purposes. By 1795, a British garrison was established and the territory became a British possession in 1814. At about this time, the black peoples (Bantu) came down from northern areas which are now being preserved for them as the basis of the homelands where they are being encouraged to live and develop their identity. Within these homelands, most of these peoples have their own governments and legislatures and enjoy a substantial and increasing measure of self-governement.

Until about 1850, the country was an almost entirely agricultural community. But in 1867, diamonds were discovered near Hopetown, a small agricultural settlement on the Orange River. It is said that children picked up a few odd specimens lying on the ground; these attracted no particular attention at first. When these odd stones were later identified, a diamond rush began, to wash gravel on the Vaal River, near the town of Pneil. The rush spread from here, up and down the Vaal and Orange rivers. Figure 6.1 shows the locations.

Until this time, diamond production had been confined to India and Brazil. Nevertheless, by 1872, the quantity and quality of diamonds produced in South Africa overshadowed all other centers of world production.

There were two types of devices for recovering diamonds from the river gravels, both rather similar to those used for placer gold in California. In both types, a series of woven wire screens of graded mesh size were used. The gravel was fed into the unit and washed through with a stream of water. The fine sand and gravel were discharged through the outer or lower fine screen. Diamonds were caught on or between the screens.

The cradle type was mounted on rockers on which it was vigorously rocked during the washing process. The cylindrical trommel type was continuously revolved, for a similar result. In each case, the screens were examined for diamonds, and the waste product was discarded as tailings.

By 1870, the alluvial diggings were abandoned because diamonds were discovered in a new area at Dutoitspan, about 20 miles to the southeast. Here the diamonds occurred in dry areas rather than in or near streambeds. By July 1871, three more areas were discovered near Dutoitspan: at Bulfontein farm, at Vooruitzigt farm (now known as the De Beers Mine), and on Colesberg kopje (now the famous town of Kimberley) (see figure 6.1). In these areas, diamonds were found in considerable quantities. A new diamond rush began, attracting adventurers from all over the world, similar to the gold rushes of California and Ballarat. Kimberley became a boom town with a population of 20,000. It absorbed as suburbs the areas of Bulfontein and Dutoitspan. Primitive frontier-style conditions prevailed. It was a typical mining camp in those early days.

At first, the miners believed that the diamonds occurred only in the surface gravels within a restricted area beyond which no diamonds were found. Later it was realized that these roughly circular or elliptical areas were the surface expressions (outcrops) of vertical volcanic pipes, with a virtually unlimited depth dimension. Near the surface in the oxidized zone, diamonds were found to occur in a matrix of soft yellow ground, described as raisins in a pudding. Below this zone, a dark blue volcanic material was found in the pipes. This became known as "blue ground" or kimberlite. This relatively soft blue ground was, therefore, the volcanic gangue, or matrix material, within which the diamonds sporadically occurred.

Individual miners were allowed to stake claims 30 feet square. Each claim was to be separated by a roadway, 15 feet wide, to give access to the claims. The blue ground was excavated manually by pick and shovel and then wheeled in carts or barrows, along the roadways to a "sieve," or crude washing plant, located clear of the mining area and beyond the surface area of the pipe. The material was puddled with water so as to pass through the sieves, while the diamonds were caught on the sieves (screens).

As the work in each claim progressed and the workpits became deeper, the walls of the access roads either caved in or were encroached upon. By 1872, the roadways had become unserviceable, and it was difficult to transport the mined material to the washing plants. At this time a simple rope and pulley system was used to hoist the ore in buckets above ground level, and then across other claims to the washing area. The thousands of ropes, up to 1,000 feet long, extending across all the workpits, gave the impression of a vast spiderweb, as pictures taken in 1877 show. The ropes were at first operated by hand-cranked pulleys, using black African labor who came to the fields from northern areas in search of employment. They worked under short contracts and then returned to their tribal kraals. This system was developed further by the mining companies in later years. Later, horse whims were used to haul the buckets of ore to the washing plants. By this time, all the roadways had disappeared and the Kimberley Mine was an impressive sight.

As the claims became deeper, the problem of hoisting the buckets of ore increased, rockfalls became frequent, and water accumulations provided challenges. Many miners were killed or injured by collapsing walls when the workpits reached depths of 400 feet or more.

By this time, the problems of individual miners had become insuperable. They could be solved only by the introduction of systematic organization and the provision of mechanical equipment, available only to companies who had adequate financial resources. Therefore, the big companies now took over and amalgamated a whole series of adjoining individual claims. By 1880, there were almost 70 companies operating on the field. By 1888, most of these were controlled by Cecil Rhodes. Rhodes came to South Africa from England in 1871 at the age of 18. Two years later, he returned to study at Oxford University. By 1883 he had become chairman of the De Beers Company. By 1890, the De Beers Consolidated controlled 90 percent of the diamond production in the Kimberley area, and also the Premier Diamond Mine north of Pretoria in the Transvaal (see figure 6.1).

In the late 1870s, the companies had to introduce new methods. Shafts were sunk beyond the rims of the pipes and levels were driven at intervals into the kimberlite ore below the lowest horizon of the surface workings. Regular underground stoping methods were employed, and the broken ore was hoisted to the surface through the shafts for treatment in the washing plants.

The Kimberley Mine was worked to a depth of more than 3,200 feet until 1914 when it closed. By this time, the mine had excavated 25 million tons of kimberlite and had produced about 3 tons of diamonds, actually about 14.5 million carats. The mine is now filled with water, the source of the town's water supply. A mining museum has since been established at the surface. Four other diamond mines are still in operation in the vicinity. At Dutoitspan, underground block caving methods are used to extract the kimberlite ore, and the operations are largely mechanized. Two other diamond mines still operate in the Orange Free State, and another, the Premier Mine, in the Transvaal. The latter has been worked to a depth of 3,500 feet. It covers a surface area of 78 acres.

The kimberlite ore, hoisted to the surface in the shafts, is conveyed to a beneficiation plant. Here, the ore is crushed to one and a quarter inches in size and disintegrated by puddling and washing. The fine washed material passes over belts dressed with grease and is discharged to waste. Any diamonds present, normally up to 1.25 inches in size have the unusual property of adhering to the grease, and are recovered at intervals. Approximately 30 million parts of ore are mined, crushed, and beneficiated to recover each part of diamond.

The daily production of diamonds from each plant is taken to the Central Sorting Office at Kimberley. Here the stones are sorted and graded by highly skilled, experienced personnel before being sent to market for cutting.

In 1905, the Cullinan diamond was found at the Premier Mine. It weighed 3,025 carats, or about one and one-third pounds. In 1907 it was presented to King Edward VII by the Transvaal government. At Amsterdam, it was cut into nine gems which are now part of the English crown jewels. (Precious stones, or gems, are weighed in carats. A carat is an international unit of weight equal to 200 milligrams.)

Alluvial diamonds were found also in Southwest Africa, by reclaiming the offshore seabeds. Elsewhere in Africa, such as in Ghana, Zaire, and Sierra Leone, both gem and industrial diamonds are recovered; also in Brazil, Siberia, and Borneo.

Stones that are too small to rate as gems, or with flaws or discolorations, have a significant market value as industrial diamonds. About 80 percent of the total world production of diamonds are used for industrial purposes: for drilling, cutting metal, and for grinding and polishing. Modern metal-working industries have been developed to rely on industrial diamonds for toolmaking. In case at some future time the supply of natural diamonds becomes exhausted, some companies in America and South Africa are now developing processes for the manufacture of artificial diamonds.

The cutting of raw diamonds is a highly specialized operation, carried out in Belgium, Holland, India, Israel, Switzerland, and in New York. Diamonds are now being cut with laser beams.

The contribution of the diamond industry to the early development of South Africa was in line with that of gold mining in America, Australia, and Canada. Diamond mining continued what wool had begun. It provided a great impetus to the development and modernization of the transport system, and attracted population and capital from abroad. In addition, it diversified industry, broadened the scope of opportunities for young people, and gave strength and purpose to political life.

Small gold finds had been made as early as 1864 (near Leydenberg and the Zambesi River) by Button and others who had followed up traces of ancient mining activity. Then, in 1886, George Harrison, following the spectacular gold rushes in America and Australia, was able to report a significant discovery on the Witwatersrand (the Rand), centered around Johannesburg in the Transvaal. The richness of the ore caused mining activities to be begun on an important scale.

Nevertheless, there was not much alluvial gold found in the Transvaal. The reefs carrying the gold had to be worked downward from the grass roots. There was no opportunity for an individual to make his fortune except by investment in a mining company. Therefore, from the outset gold production on the Rand was the work of the mining companies. The development of the Witwatersrand goldfields was achieved largely because of the skill, experience, expertise, and capital funds that had been developed by diamond miners in Kimberley. Although little firewood was available for steam power generation, there was plenty of coal in nearby deposits.

As mining operations proceeded, it was found that the ore grade was comparatively low. This raised the specter of questionable profitability. But by using the amalgamation process, or the chlorination process, ore containing as little as one part of gold in 30,000 could be made to pay, although the recovery was low.

A considerable technological advance occurred in 1889 when two Glasgow doctors and a chemist invented the cyanidation process. A weak solution of potassium cyanide was used to dissolve the gold particles from the finely crushed ore. The gangue material was separated by filtration and discarded as tailings. The pregnant solution was then treated with zinc shavings. The zinc replaced the gold in the solution; the gold was precipitated, smelted with borax flux, and cast into bars. Approximately 96 percent of the gold was recovered by this method.

The discovery of the cyanidation process was of immense value to the gold industry and to the state of Transvaal. Johannesburg grew from the status of a village to a population of 80,000 in 1892, to 102,000 in 1896, and currently to over a million: the greatest mining camp in the world.

By this time, however, the population of the Boer farmers was outnumbered by seven to three. The Boers reacted to the introduction of foreign immigrants to work in the gold mines and a great deal of antipathy arose, leading to the Boer War in 1899, But by this time the mines at Johannesburg had become well established. Cheap African mining labor was widely available, and it played an important part in the rapid development of the mines, although this naturally led to a slower rate of mechanization. Nearly 100,000 Africans were employed before the Boer War.

After the surface ore, mined from open cuts, became exhausted, it was necessary to sink shafts to extract ore from the reef by underground methods. The reefs dipped below the surface at an average angle of 25 degrees. Hoisting machinery was fairly primitive at first, but steam hoists were later used. Ore was crushed in stamp mills. Hand drilling was replaced by pneumatic drills. By 1920, wet drilling was introduced to allay the dust. The consequent reduction in the rate of incidence of silicosis (miners' lung disease) was reduced from 1.9 per 1,000 in 1927 to 0.8 per 1,000 in 1936. Similarly, by the introduction of safety campaigns, the accident rate for underground workers, per 1,000, was reduced from 4.64 in 1904, to 2.11 in 1934, and 1.22 in 1974.

In the early days, before 1880, there were nearly 450 companies, most of which made no profits and paid no dividends. By 1889, 19 large companies paid dividends, but only 8 of them paid in 1890. By 1892, 5 large mining finance houses had been established, all backed with the cash reserves accumulated by diamond mining. Two new mining groups joined these, 1 in 1917 and 1 in 1933, and these 7 now control most of the South African mining industry. The financial stability and the degree of technological progress supported by these 7 groups has enabled South Africa to become the world's leading mining country.

The mining industry in South Africa currently produces 75 percent of the gold in the western world, 90 percent of the platinum, 75 percent of the chrome ore, 73 percent of the manganese, and substantial quantities of most other economic minerals except petroleum and bauxite; it leads the world in the extent of its organization and levels of technology.

Europe

Following the disruption caused by the various European wars between 1618 and 1748, the output of base metals revived slowly until 1785. In the 1780s the Parys Mountain Mine in Anglesey, North Wales, was the largest producing copper mine in the world. Its ores were smelted in South Wales and in Lancashire. The main thrust of mining activity occurred in Britain, Sweden, Bohemia, and Spain, where many old mines were reopened with the use of steam pumps.

The recovery of silver from copper ores by liquation and cupellation with lead, developed before 2500 B.C. and reintroduced by Funcken in 1451, was still carried out; but the Pattinson process was patented in 1833. In 1850 Percival Norman Johnson introduced this process which was invented by Hugh Lee Pattinson. It revolutionized the recovery and extraction of silver from silver-lead bullion by the progressive crystallization of lead from the surface of molten bullion in large open vessels. As the crystals cooled, they were skimmed off and the bath eventually contained molten silver.

By 1850, Alexander Parkes also patented a process for the recovery of gold and silver from lead bullion. A small amount of zinc metal was added to the bath of molten bullion which was heated to the melting point of zinc. The molten zinc was then found to dissolve the gold and silver and to form a scum at the surface. This scum was skimmed off and distilled in retorts to recover the zinc and the precious metals separately. By 1870, the Parkes process had superseded the Pattinson process.

Meanwhile, copper, lead, zinc, tin, and mercury were all being produced from their ores by smelting with charcoal or coal (later, coke). In Spain, the famous Rio Tinto Mine, worked originally by the Phoenicians, produced copper and pyrite flotation concentrates from the copper ore. It is one of the largest bodies of sulphide ore in the world. This mine, regarded by the Spanish government since 1556 as its only hope of regaining national solvency, had been notoriously mismanaged since that time, mainly because of lack of capital and the dead hand of government red tape. During only two short periods had even a working profit been made and then only by the cementation process of precipitating copper from the mine waters with scrap iron. At length, in 1873, the government decided to grant a freehold transfer of its jealously guarded (and badly operated) property to foreign interests.

The London-based Rio Tinto Company then took over, built a railway to the port of Huelva, and developed the mine on a profit-earning basis, to the great advantage of Spanish citizens. It became the largest single employer of labor in Spain. The Spanish government purchased a controlling interest in this famous mine in 1955. It is now being run efficiently by the Union Explosivos Rio Tinto Company.

The Rio Tinto Mine is probably the best known copper mine in the world, and apart from those in Cyprus, the oldest mine still in operation. The mine is the largest single source of copper in Europe. Other European mines producing copper and pyrite were the Tharsis Copper and Sulphur Company (Spain), the Orkla Company (Norway), and the American-owned Cyprus Mines Corporation (Cyprus).

Similarly, the ancient mercury deposits at Almaden, in operation for over 2,000 years, are the world's largest, although mercury mines at Monte Amiata, Italy, and at Idria, Yugoslavia, are also important producers (see figure 9.1).

In 1900, the famous LKAB Iron Mine at Kiruna, northern Sweden, was discovered and set in production. This is the largest underground iron ore mine in the world. Some of its achievements and installations are famous. Other mines were opened at Malmberget, and at Strossa, Dannemora, and Grangesberg, in central Sweden. The Boliden Copper Mine was opened in 1925.

In the 1920s, the Outokumpu Mine containing copper, lead, and zinc in Finland was opened. Discovered in 1910, it started a significant series of discoveries of base metal ores in Finland. Nickel ores were also being mined at Petsamo during this period.

The Soviet Union, following the revolution in 1917, soon recognized the national importance of mineral production. They opened up large iron ore deposits in the Krivoi Rog, the Donetz Basin, and in the Urals. Coal was similarly exploited in the Kuznetz and Donetz basins, and also in Spitzbergen. Manganese deposits were exploited in the Ukraine and in the Ural Mountains, which extend southerly for a thousand miles from the Arctic Ocean to the Caspian Sea. They represent one of the famous metallogenetic provinces of the world. Copper ores were similarly developed in the Urals and in the Caucasus and central Asian states. Precious metals, including platinum, were continously mined in the Urals and in Siberia.

In Cornwall, by 1862, there were 340 mines operating hundreds of shafts and many thousands of miles of underground workings. Cornishmen had produced copper and tin to a value of 200 million pounds sterling, using hand-drilling methods in very hard rocks. These "Cousin Jacks" came to be regarded as the world's best hard rock miners. They are reported to have taken their tin-mining expertise and techniques to Germany in 1241. Yet it must be recognized that the German miners returned the compliment by bringing to Cornwall, in the reign of Queen Elizabeth, their capabilities and expertise in the treatment of copper, iron, and tin smelting.

Cornishmen are known everywhere for their reputations as skilled miners. When mining activities waned in Cornwall after the 19th century, Cornish miners migrated to mineral districts all over the world, and thereby made considerable contributions to other countries.

By the 1880s, compressed-air-powered rockdrills were introduced to the Cornish mines. Tallow candles still fulfilled the lighting needs until the 1900s when acetylene (carbide) lamps were put into use (see figure 12.5).

One of the mines, the Levant, was worked beneath the sea for a distance of 700 feet from the shore. As the mines approached a depth of 1,000 feet, the grade of copper ore was found to be very low. This was a setback until it was found that below this depth the grade of tin in the ore had perceptibly increased. A new surge of activity then developed.

But the discovery of easily won tin in Southeast Asia caused the closure of many tin mines in Cornwall. Mine taxation was heavy. Only two Cornish mines survived: Geevor and South Crofty. The last steam pump engine ceased working at the Robinson Shaft of the South Crofty Mine, after 101 years of active service,

The chief iron-producing country was Britain. In 1877, it produced 16,692,800 tons from deposits in Devon, Cornwall, Somerset, Gloucestershire, Lancashire, Cumberland, Lincolnshire, Yorkshire, Northants, and the Forest of Dean, as well as from the Irish counties of Antrim, Donegal, and Londonderry. By 1850, Britain was producing 2.5 million tons of all forms of iron annually.

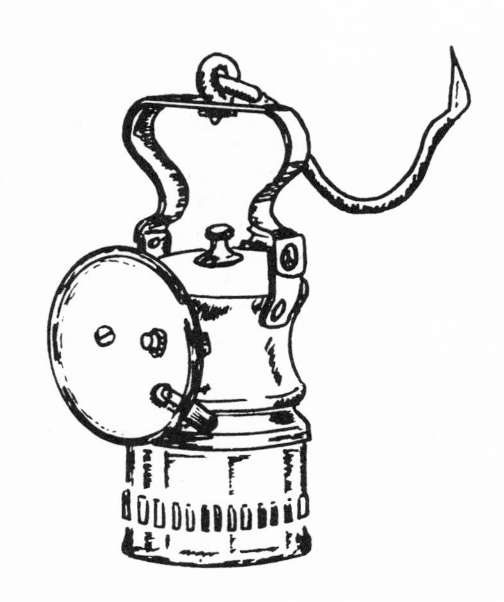

Fig. 12.5. A miner's acetylene lamp.

In 1856-57, Bessemer patented the steelmaking process, and Siemens invented the open hearth furnace. In 1875, Thomas and Gilchrist eliminated phosphorus by introducing a basic lining to the furnaces. In this way, a basic slag was formed; it had a ready sale as a phosphatic fertilizer.

Up to 1850, Britain was the largest producer of copper, tin, and lead, but after this time the industry declined rapidly owing to the discovery of rich ores in North America and to the exodus of many Cornish miners to California.

Mining commenced in the famous German potash mines at Stassfurt, near Magdeburg, in 1861. These deposits were first discovered in 810 A.D.

Asia

Coal had been mined in India since 1814 at the Raniganj coalfield. Rapid expansion of production took place between 1880 and 1900 following the development of Indian railroads (see figure 4.14). The iron ore deposits of Bihar and Orissa were first exploited after 1904, and reserves totalling many billions of tons have been developed. Manganese and chromium mining were established in 1891 and 1901, respectively.

The ancient gold mines were reestablished in the Kolar goldfield in eastern Mysore in 1884. Some of these mines were worked to a vertical depth of 10,000 feet. Old copper workings at Singbhum were reopened in 1908.

India possessed some interesting pegmatite deposits. From these, large sheets and blocks of mica were mined from 1884, and beryl was also produced. Beach sands mined at Travancore since 1912 contain ilmenite, rutile, monazite, garnet, and zircon. An aluminum smelter was set up in 1944 to recover metal ingots from the bauxite deposits first mined in 1908.

In Burma, the famous Bawdwin Mine containing silver, lead, zinc, and copper, which had been worked by the Chinese from 1412, was reopened by a British company in 1902, but it ceased operations during World War II, and has since been nationalized. Tin and tungsten mines were established after 1910.

For many thousands of years, the diamonds, rubies, and sapphires from India and Sri Lanka (Ceylon), and the Mogok rubies and jadeite from Burma have been highly prized throughout the world.

The chief mineral mined in China during the Petroleum Age was coal; but large deposits of base metals and industrial minerals were to be developed after World War II. Japan produced limited quantities of coal, iron, sulphur, and copper, but by no means sufficient to sustain her domestic demand.

In the Philippines, gold mines were established at Benguet in 1903, and at Baguio. Iron ore, manganese, copper, mercury, zinc, and chromite deposits were also developed. During the Petroleum Age, prodigious quantities of tin ore from alluvial deposits in Malaysia, Thailand, and Indonesia were mined. Malaysia still produces much of the world's tin, and also significant quantities of bauxite and iron ore.

South America

Before the arrival of the Conquistadores in the 1520s, the mineral industry of South America was based on precious metals and gems. The Spaniards sponsored the mining of placer deposits and of vein mining of gold and silver deposits, mostly in the Cordilleran region of Colombia, Peru, Bolivia, and Chile (see figure 10.4).

The earliest diamonds were mined at Tejuco (now Diamantina), Brazil, in 1725. The Bahia Mine became famous for the quality of its production. Emeralds had been mined in Colombia from before the arrival of the Spaniards, who started their own production in 1538.

Large iron ores deposits were developed in Brazil, Chile, and Venezuela by American interests in the 1940s. A tungsten deposit was also developed in Brazil. In the early part of the 20th century, large-scale production of tin commenced in Bolivia, and of nitrate in Chile. The Chilean nitrate deposits were discovered in 1830 in Bolivian territory. These were gained by Chile in 1881 after a war was fought against Peru and Bolivia. Production was highly taxed: at one time this tax yielded 80 percent of the total Chilean annual revenue. By 1900, 1,350,000 tons of sodium nitrate were being produced annually.

Similarly, the great Chilean copper deposits were developed at Braden (El Teniente) in 1912, at Chuquicamata in 1916, and at Potrerillos in 1927. Chuquicamata is the highest grade, large porphyry copper deposit in the world, averaging about 2 percent as delivered to the mill. The Chuquicamata and El Teniente deposits are two of the largest known porphyry copper deposits. Furthermore, Chuquicamata is supposed to be the world's largest single copper-producing mine. It produced 416,000 tons of copper metal in 1976. Between 1850 and 1880, Chile was the largest copper-producing country in the world.

Peru is also a famous mineral-producing state. Its reputation began with the fabulous copper-silver-gold production in 1630 at Cerro de Pasco: 550 million dollars' worth in that year. It became a large base metal mine in 1915. The largest copper mine in Peru is the Toquepala, currently producing 45,000 tons of ore per day. Others are under development.

Other Areas

Many significant mineral discoveries were made during this period in other parts of the world. The famous Copper Belt straddling the Zambia-Zaire border was rediscovered and placed in production. Copper had been known to exist in this belt since the Middle Ages. The first of the recent claims was staked in the 1890s. In Zaire (previously known as the Belgian Congo), the Belgian company, Union Minière de Haut Katanga, had worked their deposits since 1906. In Zambia (then Northern Rhodesia), the Anglo-American Co. and the Roan Selection Trust developed these properties in the 1920s (see figure 6.1).

Following revaluation of the gold price in the Great Depression, gold mines were opened in Fiji and New Guinea. In 1928, the fabulous Edie Creek placer discoveries in New Guinea started a wild rush. Claims were staked all along the creek and along the terraced banks. Coarse gold and nuggets were easily found. The finest of the gold particles had been washed down to the broad Bulolo Valley where ten dredges were placed in commission by Bulolo Gold Dredging Ltd. The center of the goldfields was at Wau. No road access was available from the port of Salamaua. All cargo, including machinery (but not timber) had to be brought in by air. Special freight carriers were designed, mostly Junkers single- and tri-motored aircraft. This was the first pioneering effort at air cargo transport.

In the 19th century, there was significant mining activity in New Zealand. A gold rush occurred in Gabriel's Gully, near Otago in the South Island, where Gabriel Read found rich deposits of placer gold in 1861. This was followed by more gold discoveries near Nelson and on the west coast. It was here that the first gold dredge in the world was built. Coal is also mined in this area. In 1867, a rush developed to the Coromandel Peninsula of the North Island, to what became known as the Thames Goldfield. The Caledonian Mine produced 330,000 ounces of gold in 1871. But a larger underground gold mine operated here between 1887 and 1914. This was the famous Waihi Mine, discovered in 1878. Since that time, mining activity has waned in New Zealand, due to some extent to political and community attitudes.

Copper deposits were developed in Cuba in 1853. A gold mine was worked on Aruba Island in 1875. Nickel ores were first discovered in New Caledonia in 1863. Large-scale mining was commenced by open cut methods in 1890, and smelters were established. The lateritic ores cover much of the surface of the land. The average grade of ore mined was 9 percent nickel in 1890, and about 2.5 percent today.

13 The Uranium Age (from 1950)

World War II came to an abrupt end in 1945 following the use of two atomic bombs released on two Japanese cities. The effect of this weapon was tremendous. But the basic raw material was one of the complex uranium minerals.

These minerals had previously been obtained from Joachimsthal in Bohemia, and were used by Mme Curie in 1898 to isolate the elements: radium, thorium, and polonium. Some radioactive ores had also been produced from the Belgian Congo (now Zaire). Small deposits had been found in 1881 near Uravan, in Colorado, and in South Australia.

The end of World War II created a tremendous impetus to secure supplies of uranium ore to maintain Allied supremacy in the atomic weapons field. Markets and prices were guaranteed. The "uranium rush" began in 1949 in New Mexico. All over the western part of the United States people were prospecting with Geiger counters. This was an electronic instrument designed to detect the presence of radioactive materials. By 1951, radiation signals were being received by instruments carried in airplanes. Many deposits were found in Colorado, Utah, Oregon, Washington, Wyoming, and New Mexico.

Mining of uranium ore in the United States grew from 15 established mines in 1946 to 1,000 mines in 1960, yielding a total annual production of 7 million tons. By 1972, America's estimated reserves rose to 97 million tons.

But uranium ores were also being mined in other countries, such as the Belgian Congo. In the milling of many South African gold ores, uranium oxide was now being recovered as a by-product. The Blind River Mine was opened in Canada. Throughout that country, more than 1,000 finds were reported by 1956 and Canada was producing 37 percent of world output by 1959. The Rum Jungle Mine (found in

1949) and the Mary Kathleen Mine (discovered in 1954) produced significant quantities of uranium "yellow cake" in Australia, as did the Radium Hill and the South Alligator mines. Within ten years, Australia exported more than 164 million dollars' worth of uranium concentrate for a total capital investment of 50 million dollars. But by 1956, following the withdrawal of exploration incentives, a slump set in. Nevertheless, many stupendous uranium finds have since been made in Australia. In the Northern Territory, Australia has some of the world's richest uranium deposits. It now holds more than one-fifth of the world's uranium resources.

From 1950 onward, advanced mechanization programs were set up in coal mining, and more particularly in longwall mining. Although pioneered many years earlier in Britain and Europe, the first mechanized longwall unit installed in the United States was in 1960. By 1973, there were 65 units operating, thereby demonstrating a trend away from the less efficient room-and-pillar method. Mechanical loading of coal had begun in the United States in 1925. By 1973, 60 percent of all coal mined underground was won by continuous miners (room-and-pillar method), 35 percent by conventional room-and-pillar mining, and only 4 percent by longwall mining.

At Bevercotes Colliery in 1963 in England, a partly automated longwall system was under trial. It involved a sequence of mechanized operations at the face without the presence of workmen. This system was known as ROLF, an acronym for Remotely Operated Longwall Faces. It involved remote control and instrumentation of the hydraulic roof supports, the coal-cutting machine, the face conveyor, and other units and services to mine coal continuously. The control console was located in the main roadway, back from the coal face. Various sensors fed signals to the control panel.

At a given signal, each roof support unit extended a horizontal ram to push the conveyor to the new coal face. It then retracted from the roof, pulled itself forward (using the conveyor as an anchor block), and reset itself to the freshly exposed roof. Each roof support unit performed these operations in a programmed sequence, following the passage of the cutting machine along the face. When the latter reached the end of the longwall face, the direction of operations was reversed. All operations were monitored by one man at the control console. Otherwise, the only workers present were maintenance men.

In the following year, ROLF II was installed. This was an improved model. The results of these pilot schemes were encouraging. Productivity was higher. In one week, ROLF II, under full production conditions, produced 23 tons per manshift, as against a national average of 6.5. Approximately 15 percent of all longwall faces presently operated in Britain use automated procedures.

Other developments in coal-mining practice during the Uranium Age include the shortwall method, introduced in Australia and now being tested in other countries including America. This is a hybrid method, designed to incorporate the best features of both the room-and-pillar and the longwall methods. Both continuous miners and hydraulic-powered supports are used. This method is expected to provide greater protection from roof falls and improvement in coal output. It can also be adapted to extraction of coal pillars.

Following the depletion of high-grade iron ores in the Lake Superior region of the United States in the 1940s, it was realized that other similar deposits would have to be found. These were discovered in Labrador and Quebec (see figure 12.4). Other deposits were developed in Venezuela and Brazil. Meanwhile, attempts were continuing to develop uses for the low-grade (30 percent Fe) iron ore still remaining in the upper Minnesota region. In order to beneficiate these to a higher grade, say 60 percent, it was necessary to crush and grind the run-of-mine ore. But this produced a size of particle that was physically unsuitable as blast furnace feed. So a new process was then devised. These fine particles are now "pelletized." The use of high-grade pellets of iron are in general use and the mining of low-grade ore is proceeding at a regular pace.

A large, new magnetite ore body, buried well beneath the surface, was discovered in Missouri by drilling an anomaly plotted from aerial magnetometer surveys. This Pea Ridge Mine began production in 1964.

Following World War II, Canada entered its greatest period of mineral expansion. Existing mines increased production, and new scientific exploration techniques (developed during the war and adapted to mineral prospecting) brought in a number of new producing mines. Among these were the great iron ore deposits in the Quebec-Labrador region; the large uranium discoveries in Ontario and Saskatchewan; a second major nickel complex in Manitoba; the vast potash deposits in Saskatchewan; a large lead-zinc mine at Pine Point in the Northwest Territories in 1951; and many other base metal mines across the country.

The value of all types of minerals produced in Canada escalated over 1,000 percent from 1945 to the early 1970s. The growth in mineral production was closely interwoven with Canada's industrial development.

By a wise national policy of encouragement of mineral exploitation through fiscal incentives over many years, Canada has developed its mining capacity to the point where it is a truly national industry. Canada leads the world in annual mineral production value per capita. The momentum thus gained represents a real national asset. But any reversal of this policy is bound to dampen mineral production incentives.

During the last three years Canada appears to have killed its goose. There is no better way to stop or slow the supply of golden eggs. It is very refreshing to note, however, that the goose still breathes in British Columbia. The minister of mines is reported (CIM Bulletin, August 1976) to have introduced two bills. One

> . . . designed to restore the confidence of investors and resource companies in mining in the province and also to provide jobs for thousands of British Columbians.

And another to restore

> . . . jobs that were lost as a result of the punitive and repressive measures contained in legislation enacted by the former government.

Another metal that gained a level of world importance in the Uranium Age, following World War II, was aluminum. Although aluminum is the third most abundant element in the earth's crust (8 percent), it was not isolated as a metal until about 1845, and a commercially successful method of production from its ores was not developed until 1886. At this point, aluminum metal was a novelty and had few industrial uses.

Following World War II, the development of aluminum metal for cladding aircraft fuselages received a tremendous impetus. As a result, aluminum is now second only to iron as the most strategic commodity in our daily lives.

The ore from which aluminum metal is recovered is bauxite. Apart from small deposits in Europe, most of the world's bauxite is now mined in Australia, in the Caribbean countries, and in Malaysia.

For years Australia imported all its bauxite from Malaysia and Indonesia for the aluminum smelter established in Tasmania in 1955. But by 1964, Australia began proving enormous deposits of its own at Weipa in Cape York Peninsula, in the Northern Territory, and in Western Australia. That country now controls over 25 percent of the world's reserves, and exports large tonnages of bauxite and alumina. Australia also has developed the enormous low-grade copper-gold deposits on Bougainville, in the Solomon Islands. Another development is beach sand mining. Heavy minerals such as ilmenite, rutile, zircon, and associated minerals are mined in beach deposits in Australia. This is the source of most of the world's rutile.

During the 1960s a tremendous upsurge in exploration activity occurred in Australia. Apart from the bauxite and uranium discoveries, huge deposits of high-grade iron ore, manganese, nickel, phosphate rock, and coking coal were developed; large reservoirs of petroleum and natural gas were brought into production; and five large diamond pipes were discovered, logically enough in the Kimberley Ranges. The tempo of mineral exploration activity in Australia during this decade was more intense than any previously experienced in world history. The momentum generated attracted large blocks of capital investment from abroad, partly because the political climate for investment was so beneficial.

However, in 1972, as in Canada, a new immature government adopted foolish measures in a spirit of local nationalism that soon killed the golden goose. Exploration activity virtually ground to a halt. Although this government was replaced at a general election in 1975, it will take many years to recover the exploration momentum so imprudently destroyed.

The old lead and zinc mines of Zawar were reopened by the Hindustani Zinc Corporation. The ore is smelted at Debari, near Udaipur, India.

Although Scythian gold has been recovered from the Altai Mountains in Siberia since about 500 B.C., and alluvial mining was producing one-quarter of the world's output of gold by 1884, the mineral development of Siberia accelerated after World War II. Despite extremely severe climatic conditions, immense reserves of copper, nickel, gold, asbestos, tin, iron, lead-zinc, diamonds, coal, petroleum, and natural gas have since been developed; the completion of the new Baikal-Amur railway is likely to add significantly to these reserves. Siberia must be rated as one of the world's fabulous storehouses of mineral wealth.

Since World War II, there has been a growing demand for metals, fuels, and mineral raw materials to supply the industrialized countries at increasingly larger levels of consumption. There have also been moves to assist the growing number of Third World countries who are seeking to achieve a measure of industrialization. With few exceptions, this will hardly be possible unless a particular country possesses a developed supply of mineral raw materials both as a base for a new domestic industry, or to export for foreign exchange. This movement is already taking place in Middle East countries where oil production is their only significant source of mineral wealth.

Since 1950, the major sources of the main mineral commodities have swung from one country to another during the development of recent large exploration campaigns. It is interesting to follow these developments.

Also, the use of scientific principles and machines during the two world wars has been applied increasingly to the technological branches of industry. Many of these sophisticated technical developments have been incorporated into the search for and development of mineral deposits, in the mining of minerals, and in higher technical grades of products from the run-of-mine materials.

III

Chronological Development of Particular Aspects of Mining

An Introduction to Part III

The development of mining methods and systems was dealt with in part 1 in a necessarily simplistic way; in part 2 particular mining operations were briefly described. The material in part 3 deals with specific aspects of procedures, ancillary to the actual extraction of minerals, where these have evolved as part and parcel of the gradual development of technology in mineral production.

14 Mine Drainage

Beneath the surface of the earth's crust there is an equilibrium level of water saturation. Below this level, known generally as the ground water level, all open spaces are saturated with water. The equilibrium condition is maintained by the seepage of rainfall to balance the loss to lakes and oceans. The level is generally higher in areas of high rainfall. When mining activities progress below ground water level, water will continually seep into the workings. If not removed by some sort of drainage procedure, the water will naturally fill the mine to the normal ground water level.

In the early days, mining was for the most part carried out in shallow workings, probably above the ground water level; if below this level and the seepage flow was unmanageable, mining activity had to cease. Otherwise, until this critical point was reached, the removal of water from the workings represented a difficult problem.

In the mining of iron ore at Mitterberg in the Tyrol between 1600 and 400 B.C., drainage was achieved by excavating a drainage adit into the hillside below the mine workings. This effectively lowered the ground water level so that ore extraction could be carried out above the adit level without undue water seepage problems.

Although an expensive procedure in terms of first cost, this adit drainage method has since been adopted in many wet mines where the topography is suitable, more especially before the invention of pumps. Some typical mining districts where drainage adits have since been used are in Cornwall, the Harz Mountains, the Erzgebirge, Nevada (the Comstock Lode "Sutro Tunnel"), and Cripple Creek, Colorado. In the Harz Mountains in 1851, the Ernst August Adit was commenced to drain the mines at Clausthal to a depth of 1,200 feet. With branches, the total length of this completed drainage adit was

26 km. Other long drainage adits were driven in Freiberg, Saxony, and at Schemnitz. The drainage adit in Cornwall was begun in 1748. During the next 50 years it drained 46 mines. With branches, it totalled 30 miles in length, although the furthest mine was only 5.5 miles from the portal.

In Roman days, the Empire was so extensive that metal production became an important state activity. There were many mines that could not be drained by adits. In relatively shallow mines, a moderate amount of seepage water was baled out by slaves using buckets made from esparto grass soaked in tar, or from animal hides. But as the mines became deeper and the seepage flow increased, improved methods were demanded. The Romans made some important contributions to drainage methods. Two drainage devices developed by them were the water wheel and the Archimedean Screw, or cochlea.

At the Rio Tinto Mine in Spain, a water wheel system was introduced by the Romans. Water wheels were the main source of power (other than human and animal) for about 400 years. The water wheel had previously been used for developing motive power from a flow of water in rivers, such as we see today associated with old flour mills. But to drain mines, the process had to be reversed. A source of power was necessary to rotate the wheel to lift the water. Such wooden water wheels since found at the Rio Tinto Mine were about 14.5 feet in diameter. They carried 24 buckets or water boxes spaced around the periphery. These wheels were operated by slaves in treadmill fashion. Each wheel raised water about 12 feet to a higher cistern. A series of such wheels was used to raise water to the surface from a considerable depth. One such arrangement discovered at the Rio Tinto Mine in 1920 showed 8 water wheels in series, raising the water about 95 feet at the rate of 20 gallons per minute.

The other device, the cochlea, was invented by Archimedes of Syracuse who first noted these in use in Egypt for irrigation purposes on the Nile. It consisted of a wooden rotor carrying helical blades of wood or copper, rotated in a wooden barrel by a slave operating a crank or a treadmill. Some had a fixed flight of vanes, about which the cylindrical barrel was rotated. The device was set on an angle, the lower end immersed in a sump to pick up the water which was then delivered to an upper sump or cistern. A series of these devices was necessary to lift the water to the desired discharge point. Each unit lifted water only about 6 feet vertically. These cochleas were in widespread use in Roman times when some mines were worked to a depth of 600 feet; but they were expensive to install and costly in terms of manpower. They were used only in the richest of mines, or in those mines where water seepage was too great to handle by simple baling methods: as for instance, by a "bucket brigade" of men passing buckets of water from one to the other up the ladders of a shaft, or by winding them up on a hand windlass. No such device was found in Britain following the Roman occupation.

During and following the Middle Ages, advanced drainage devices were used. These were described by Agricola in his book *De Re Metallica*. The removal of mine water had been the main problem in mining. The devices described by Agricola were much superior to those operated by the Romans.

One such device was a "rag-and-chain" pump, operated by men or horses, or by a surface water wheel. Six different types of these were described by Agricola. A number of hollow metal balls were attached at intervals to a continuous chain passing over a pulley at the surface, and dipping into the sump at the bottom of the shaft. As each hollow ball dipped into the sump, it filled with water and discharged its load into a trough when it passed over the pulley at the surface. Another type worked in a wooden pipe, so that the balls, stuffed with rags, brought water to the surface as they passed upward through the pipe (see figure 14.1). Still another type used bundles of rags, not enclosed in metal balls, attached to the moving chain at intervals within the pipe.

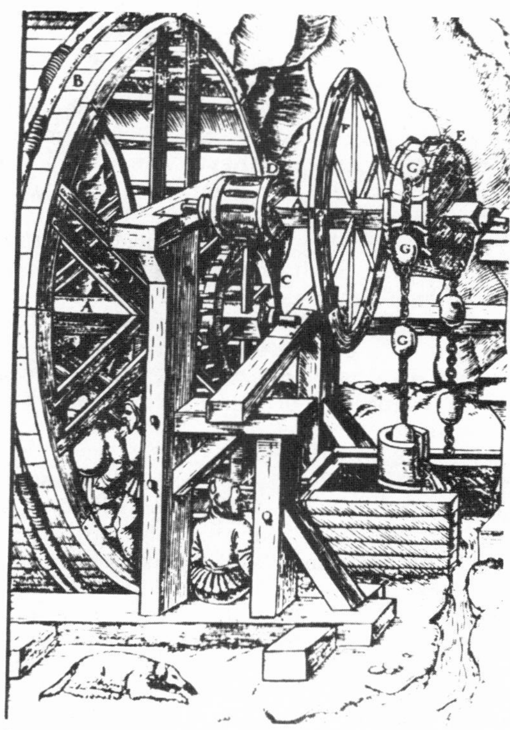

Fig. 14.1. A rag-and-chain pump.

One such rag-and-chain pump installation was driven by 32 horses in groups of eight for 4-hour shifts, with 12-hour rest periods. Another was installed in a mine near Schemnitz in the Slovakian Ore Mountains, where three such units were installed in series to operate from a depth of 600 feet. Ninety-six horses were used. They had to be led to lower levels by an inclined shaft or ramp. This was a rather unusual arrangement, however. Mines were rarely so deep at this time, seldom more than 100 feet.

Other devices had also been introduced, including those that sucked water from the mine by a series of lift pumps (pistons operating in cylinders), no doubt the forerunner of the modern plunger pump. These were operated in series by pump rods actuated by water wheels at the surface. Details can be seen in a mine map prepared by Daniel Flach in 1661. This shows the underground workings of a series of mines between Wildemann and Zellerfeld in the Harz Mountains, covering a strike length of four kilometers. The map shows the "13-fathoms Drainage Adit" and a series of hoisting and piston-pumping installations, all actuated by water wheels. At Zellerfeld in 1680, there were 6 water wheels at the surface, and 18 underground, all actuating a series of piston-type pumps through connecting rods. These lift pumps were first used at Joachimsthal in 1550 and at the Rammelsberg Mine in 1566.

As mines became deeper, however, these methods proved inadequate. Better solutions to the drainage problem were sought. In the 17th century, there was a sustained incentive to use steam as a source of power for general pumping and other purposes. But it was in coal mining that the most intensive efforts were made. David Ramsay obtained a patent in 1631 to raise water with steam power. In 1702, Thomas Savery issued a pamphlet describing a machine to raise water from the Cornish tin mines. The principle on which his patent was based involved the condensation of steam in closed vessels to produce vacuum effects and pressure differences which were translated into a sucking action. However, it was not of much practical value.

In 1712, Thomas Newcomen, a Dartmouth blacksmith who had been working independently on similar lines, developed his first pumping engine at the Dudley Castle Coal Mine in Worcestershire. It involved the use of a piston working in a vertical cylinder with steam at atmospheric pressure. The weight of the pump rods in the shaft (see later) caused the piston to rise and to draw steam into the cylinder. At the top of the stroke, cold water which was sprayed on the outside of the cylinder caused the steam to condense, forming a vacuum, and the piston was forced down by atmospheric pressure. Newcomen had to combine his efforts with Savery who held the rights to the patent. At a speed of 12 strokes per minute, this engine raised ten gallons per minute against a head of over 153 feet, through a number of pump units spaced one above the other. It had a greater capacity than 5 Savery engines. By 1769, there were 120 Newcomen engines installed at coal mines in Britain. His first engine erected in Cornwall was at the Wheal Fortune Mine. It had a 47-inch diameter steam cylinder and operated at 15 strokes per minute; it was capable of pumping for a lift of 180 feet, consuming 12 tons of coal per day. This was costly for the Cornish tin mines, but a small problem for coal mines.

Newcomen's engines were erected in Hungary, France, Belgium, Austria, and Sweden before he died in 1729. One was installed in a Newark, New Jersey, copper mine in about 1750. For 66 years at least two engines were built per year in England. This engine greatly assisted mining in Cornwall and revived deep coal mining in the north of England. It also pumped water supplies for towns and was used to feed water wheels in low-lying country.

Cornish engineers next concentrated on efforts to improve the Newcomen engine in order to reduce coal consumption. In 1767 John Smeaton began to develop more efficient engines with larger cylinders. By 1775, he had improved the Newcomen engine to give double the pumping duty. He erected a 72-inch engine at the Chasewater Mine, and pumped the mine to the 360-foot level. One of his engines was erected in Russia. The last Smeaton engine was dismantled in 1934 after operating without a serious breakdown for more than a hundred years in Yorkshire.

James Watt, a Scottish engineer, also improved on the Newcomen engine. He used direct steam pressure on the piston in conjunction with a separate condenser. Watt formed a partnership with Matthew Boulton. Watt's first two engines, built in 1776, were used to drain a coal mine and to operate a blast furnace. In 1777, a 52-inch and a 30-inch engine in the Boulton and Watt design began working in Cornwall. Both engines proved successful and reduced coal consumption significantly. By the end of 1783, 20 Boulton and Watt engines had replaced 74 of the older style engines. By 1785, they were being used in cotton mills, breweries, and other industrial plants. All told, 55 Boulton and Watt single-acting pumping engines were erected in Cornwall.

In 1779, the first steam engine was installed in Paris to pump the city water supply. About 20 Watt engines were being manufactured in the Soho works in London in that year. By 1800, nearly 500 had been built. In 1813, a Watt engine was installed on a coal mine in the Liège district of Belgium. Others were erected in Pennsylvania. The Watt and Boulton engine was later improved by Trevithick, and still later by Arthur Woolf, both Cornish engineers.

By 1820, the Cornish beam engine, basically designed by Newcomen, developed by Smeaton and then by Watt, improved by Trevithick, and further improved by Woolf, had become world famous for its simplicity, economy, and efficiency. It had enabled the Dolcoath Mine to operate from a depth of about 1,400 feet. But more importantly, this steam engine had made possible the Industrial Revolution which occurred in Britain in 1760-70.

At the Burra Burra Copper Mine in South Australia, a 104-inch engine was in operation up to 1877. This engine worked at four strokes per minute, pumping two million gallons per day from a depth of 600 feet.

The last Cornish pump engine ceased operating in 1955 at the Robinson Shaft of the South Crofty Mine, Cornwall, after 101 years of successful service. Cornish steam pumps had solved drainage problems all over the world. However, although simple in operation, they were very cumbersome. Tall stone engine-houses were required. The pump rods operating in the shaft were variously 12 x 12, 14 x 14, and up to 24 x 24 inches in section. A terrific dead weight had to be balanced and the inertia losses were considerable.

With the advent of electricity, and with much smaller multiple plunger pumps and more recently developed multistage centrifugal pumps, all driven by electric motors, tremendous improvements in mine drainage have been made. Probably the wettest mine in the world is the Bancroft Mine (now known as the Konkola Mine) in Zambia, from which over 100 million gallons per day from a depth of about 2,300 feet are pumped.

But mine drainage difficulties have no doubt been the cause of premature closing of many mines based upon otherwise profitable ore bodies. The old hands, when asked for the reason for closure, invariably state: "The water beat 'em." Such a statement can be readily understood. But how many old mines around the world would it now pay to reopen, in an era of higher metal prices, advanced technology, and certainly greatly improved pumping equipment? The question is of entrancing interest and the source of considerable speculation.

15 Underground Haulage

The question of haulage of ore or coal along the levels to the shaft, or out of the mines through adits, provides an interesting evolutionary study. In the earliest days of underground mining, as in Egypt, where the working faces extended for hundreds of feet along the vein, the gold ore was transported in panniers, presumably by slaves.

In Mitterberg, ore was carried in leather bags; at Laurium, Greece, it was carried or dragged to the shafts in baskets. Agricola writes of the use of wheelbarrows where the work faces were some considerable distance from the shaft. Some mines had a series of planks laid along the floor of the drift with a central slot between two adjacent parallel planks. Small wooden wheelless trucks were hauled along the planks, with a pin or dowel engaging loosely with the slot between the planks. In this way, the trucks were not readily "derailed." The attendant noise was much like the growl or bark of a dog, and so these wooden truck boxes were called dogs. Perhaps this term is the source of the modern dogspike for fixing rails to wooden ties in railroad practice.

In British coal mines in the 17th century, broken coal was hauled along the galleries to the shaft in wicker baskets called corves, loaded onto sleds. The sleds were kept in position by the dowel and slot method just described. Corves generally held about 500 pounds weight of coal. Later, wooden rails were used.

In the 18th century, a colliery manager named John Curr invented a wheeled corve made of wood to run on cast-iron rails to the shaft, and then to be hoisted up the shaft. This dispensed with the need for the sled and the planks. Iron rails were introduced in the coal mines at Liège, Belgium, in 1800. Where the seams were thick, pit ponies were introduced in 1763 to haul the wheeled corves, now known as trams.

In the early 19th century, especially in thin coal seams with insufficient headroom for wheeled corves, the coal conveyor was introduced. The earliest type was a low, flat car mounted on rails and moved along the longwall face as required by using an endless rope operated by a boy.

Next, the Blackett conveyor was introduced, operated by compressed air. The face conveyors fed the trams in the main roadways. These trams were hauled to the shaft on a rope system operated by a stationary steam engine. The trams were hooked or clamped to the moving rope at intervals. Endless rope haulage had been introduced by John Buddle in 1844.

Early in the 20th century, small compressed air locomotives were used to haul coal cars along the main roadways. Electric trolley and diesel locomotives are currently used. In many coal mines, main conveyor systems now carry all coal to the shaft, fed variously by face conveyors or shuttle cars.

In metal mines, rail-bound locomotive haulage to the shaft has been common since the 19th century. In small mines, steel-wheeled box-type ore cars were pushed manually on steel rails, and either discharged into an ore pocket at the shaft, or run into cages for hoisting to the surface. In parts of Germany, some mines completely mechanize this operation of caging cars.

For large mines, locomotive haulage is the rule. A connected rake of ore cars is hauled to the shaft and discharged into the ore pocket without uncoupling. Such devices include the Granby system, the rotary dump system, and the Swedish train system.

The first electric locomotive ever successfully operated now rests in the Technological Museum at Munich. It was introduced into underground mining service in 1883. It is interesting to note that the first practical electric streetcar was operated on the Frankfurt-Offenbach circuit with overhead trolley supply in 1884. Similar services were installed in Richmond, Virginia, in 1888 and in Denver, Colorado, in 1889. An electric locomotive replaced ponies in the Greenside Mine in the Keswick area of Britain in 1890, and gave good service for 40 years. Later examples of early mine locomotives may be seen at the Copper Queen Mining Museum in Bisbee, Arizona.

The first steam locomotive was built in 1801 by Richard Trevithick, a Cornish mining engineer, for coal mine operation. The first surface railway was erected on a coal mine in 1804; an exhibit occurs in the Railway Museum in York. In 1806, a tramway was surveyed from Dolcoath to Portreath Harbour in Cornwall but never constructed. A later tramway was built to serve other mines in Cornwall. It was operated by horses. In 1814, George Stephenson, a mining engineer, constructed a steam locomotive to operate at the Killingworth Colliery. In 1824, a 4-foot gauge railway using a steam locomotive carried over 50,000 tons of ore and 20,000 tons of coal. In those days, these mines yielded more than one-third of the world's output of copper.

It was in 1825 that George Stephenson built the first passenger railway to run between Stockton and Darlington, using a steam locomotive called Locomotion. By 1830 he had built the Rocket to operate on the Liverpool and Manchester Railway. This famous engine became the prototype of steam locomotive design in the years to come. It is interesting to note that George M. Pullman invented the pullman sleeping berth for passenger trains, based upon his experience with miners' bunks in the mine bunkhouse accommodation at Leadville, Colorado.

Hydraulic transport of coal or ore in pipelines is currently being considered as a means of underground haulage.

16 Hoisting

Where the access to an underground mine is by a shaft sunk from the surface, it is necessary to hoist the broken ore or coal up the shaft. In some cases, the haulage cars are placed directly in cages and hoisted; sometimes the ore is drawn from ore pockets into special ore skips and hoisted up the shaft, and automatically dumped into a bin on the surface. Meanwhile, the miners and all sorts of supplies and machinery must be hoisted up and down the shaft in cages. But of course much simpler and cruder methods were used in the early days of mining before the development of motive power.

At first, wicker baskets or leather bags of ore were hoisted up the shaft manually by means of a rope. Later a simple man-operated windlass was used (see figure 11.1). Miners, too, were lowered and hoisted by windlass, either in a basket or standing in the loops of a rope. For shallow shafts, some mines were equipped with tree ladders, formed by trimming the branches from a tree trunk. Runged ladders were later used, with a stage support between individual ladders, inclined at an angle.

Later types of hoists were the whip and the whim, both operated by horses. The whip employed the straight-line pull of the rope over two pulleys (see figure 11.1); the whim was a device using a rope drum mounted on a vertical spindle near the shaft. Two horses operated the whim carousel-fashion from extended spars. The ropes passed over pulleys mounted over the shaft (see figure 16.1). As the horses walked around the circular path, the ropes were wound onto or unwound from the vertical drum. In the Harz Mountains during the 16th century, kibbles (made of wood, bound with iron, and holding up to 300 kg of ore) were hoisted in balance by a horse whim with as many as four horses, or by a water wheel.

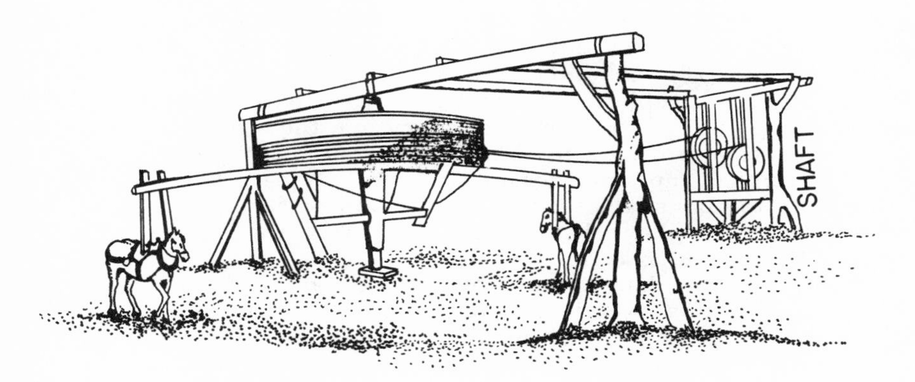

Fig. 16.1. A whim hoist.

For deep mines, hoisting chains were used instead of hemp rope, as at the Rammelsberg Mine in 1565. These presented a serious problem because of the great weight of the chains. Horses could not lift a load of ore or coal to the surface in a single lift. Two or three stages of hoisting were required. In some mines, as in Germany, water wheel power was used.

The problem was solved in 1784 when Boulton and Watt developed a steam hoisting engine. By 1850, these were in common use. But there were still two further problems to be solved. Corves became twisted in the shaft, thereby fouling the shaft timbers or colliding as they passed one another, when operating in balanced pairs. This was solved by introducing iron cages running in shaft guides. Wheeled corves were then pushed into these cages for hoisting. Secondly, the weight of the hoist chains represented too large a part of the total load to be hoisted. At Dolcoath in Cornwall, in 1784, a horse whim was able to hoist 330 pounds of ore from the 540-foot level, but the hoist chain weighed more than one ton. Shortly afterward, wire ropes were imported form Germany and this problem was overcome. Wire ropes had been invented in the Harz Mountains of Germany in 1833.

Developments in steam hoisting became spectacular later in the 19th century and in the early 20th century. Considerable improvements were made in rope manufacture, in engine design, and in the use of special shapes of hoisting drums. Probably the outstanding museum piece is the hoist room of the old Quincy Shaft at Houghton, Michigan. Here, a mammoth steam hoist, using bi-cylindro-conical drums 30 feet in diameter, was in service until 1930.

The hoisting of men provided an interesting phase of mining development. When men had to climb over 1,000 feet of vertical ladders at the end of a shift, the Royal Cornwall Polytechnic Society in 1841 offered a prize for the best solution to the problem. A machine was derived from the practice adopted by some miners of riding the Cornish pump rods. Cleats were nailed to the wooden rods at intervals. As the rods stroked up and down for 6 to 12 feet, the miners stepped off to the shaft timbers at the top of the stroke, and stepped onto a higher cleat at the bottom of the stroke. In this way they could ride up the shaft.

But a machine had already been developed at Clausthal in 1833 to imitate this practice, in a shaft 220 meters deep. In 1835, a second engine was installed to operate to a depth of 430 meters. One was installed in Cornwall in 1842 at the Tresavean Mine. The "man-engine" actuated two vertical rods operating in opposite directions and extending down the shaft. Small platforms were attached to the rods at suitable intervals. At each stroke, a man could be raised 12 feet by stepping alternately from one rod to the other. By 1900, these machines fell into disuse, however, because they were replaced by cages. But they had been used by mines all over the world. The writer saw such a man-engine in the Samson Shaft of the old St. Andreasberg Mine (1528-1910); it had been powered by a wooden water wheel. This old mine is now retained as a museum, in an attractive wooden setting in the Harz Mountains. It is reported that another man-engine was erected at the Maria Shaft at Prziban, Czechoslovakia, to replace an older model installed in 1867. The Maria Shaft operates to a total depth exceeding 3,000 feet.

Most steam hoists used drums on which the ropes were alternately stored and paid out as the conveyance (cage or skip) was hoisted up or down its shaft compartment. Each hoist had two drums to accommodate a two-rope balanced system, oerating in two adjoining shaft compartments.

In about 1860, however, the Koepe friction hoist was introduced in Europe. This system had special advantages, as typified by the 11-hoist headframe installation at the LKAB Iron Mine at Kiruna, Sweden. It was not until a hundred years later that the first friction hoist was installed in America.

All manner of improvements have recently been incorporated in hoist design. Since about 1920, most hoists have been electrically driven, utilizing both a.c. and d.c. systems with electronic controls and safety devices, also with regenerating systems. Some of the hoists installed in South Africa are of 6,000 horsepower.

17 Hard-Rock Breaking

In early times where the rock or ore was too hard to release from the face with hand tools, such as the hammer and gad, or the pick, it was first loosened by a process known as "fire-setting." A fierce brushwood fire would be set against the face. The heat generated was used to loosen the individual minerals by the principle of differential thermal expansion, aided by contraction as water was later thrown onto the heated face. However, fire-setting vitiated the atmosphere, especially where sulphide minerals were present. Nevertheless, fire-setting was used in some European mines until late in the 19th century: up to 1878 in the Rammelsberg Mine.

Fire-setting was gradually replaced by the use of explosives. Blasting powder was first used in Schemnitz, an old mining town in the Slovakian Ore Mountains between present-day Bratislava and Kosice, in 1627, in the Oberbiederstollen Mine, by Kaspar Weindl, a miner from Tyrol. Although the use of gunpowder for mining purposes in Freiberg was proposed by Martin Weigal in 1613, it was not in general use in the Harz and the Erzgebirge until 1630. Actually, it was introduced to the mines of Clausthal in the Harz Mountains in 1632, and to the Hohe Birke Mine at Freiberg by the Clausthal miner, Kaspar Morgenstern, in 1643. In about 1638, German miners introduced gunpowder for underground blasting at the Ecton Copper Mine in England. It was in general use in Cornish mines by 1689.

Later it was realized that the use of a pile of gunpowder against the face was both wasteful and ineffective. It was necessary to place the explosive within the rock mass behind the face to burst the rock outward. To achieve this, it was necessary to drill holes into the face in strategic positions to accommodate the explosive, which was then sealed in the hole by clay stemming material.

At first, the charges in these holes were ignited by using a straw filled with gunpowder. One end of this straw was ignited by a match, and the straw burned gradually toward the charge while the miner withdrew to a place of safety. This was improved by William Bickford who invented a "safety fuse" in 1831.

The drilling of blastholes then became an important skill in the repertoire of the miner. Until the late 19th century, holes were drilled in hard rock by successive blows on a chisel-pointed drill steel with a heavy hammer. Between blows the steel was rotated a little so as to maintain the circularity of the hole. Chippings were removed with a scraper. Holes were generally up to three inches in diameter and three to four feet deep, drilled in any required direction, either by one man or a team of two or three. Great teamwork was required to sustain morale, especially among unskillful wielders of hammers and nervous holders of steel. The bit of the steel was kept reasonably sharp by reforging it at intervals.

Because of the skill required and the arduous nature of the work, serious efforts were made in the second half of the 19th century to mechanize this operation. As early as 1803, a machine was developed at Salzburg, but it was apparently unsuccessful in practice. Then, Richard Trevithick of Cornwall developed a rotary hand-drilling machine in 1813. Between 1850 and 1875, about 60 patents were filed for drilling machines in England.

The next machines developed sought to imitate the manual actions of striking the back end of the drill steel with a hammer, i.e., a percussive action. Early types were developed by the Singer brothers (1838), by Brunton (1844), and by Pidding (1853). The prototype of the Schumann drill, developed in 1853, now reposes in the museum of the Freiberg School of Mines.

A steam-driven machine was patented in 1849 by J. J. Couch of Philadelphia, and another by an engineer named Bartlett, in August 1855, for use in driving the Mont Cenis tunnel through the French alps. A later machine, operated by compressed air, was developed by Sommeille in 1867 for use in this tunnel, with a greater degree of success. Fontainmoreau had introduced a drill operated by compressed air in January 1855. Later models were developed by Edward Crease (1862), Jordan and Darlington (1866), and Osterkamp (1870).

In 1871, Simon Ingersoll and the Rand brothers developed what was known as the piston drill, also operated by compressed air. In this machine, the drill steel was clamped to the piston and it therefore reciprocated with it. It had a stroke of 4 to 6 inches, and delivered from 200 to 500 blows per minute. This was a heavy machine that required the strength of two men to handle it. Similar models were introduced in 1875 by Beaumont, by Burleigh, and by McKean. Such a machine was used in the Rammelsberg Mine in 1876.

These early piston drills were later superseded by the lighter, faster hammer drills in which the drill steel did not reciprocate but remained approximately stationary, with its bit in contact with the rock face, rebounding slightly and being rotated a little between successive blows. The rear end of the steel was struck by the light, high-speed alloy steel piston.

This hammer drill was patented by J. G. Leyner of Denver in 1897. It also used hollow drill steel through which a jet of air and water could be forced to remove the rock cuttings. In this way, the dustiness of the drilling operation and hence the incidence of silicosis (miners' lung disease) were almost completely avoided. These "wet" drills were in general use throughout the world by 1920.

In the years since that time, great advances in rock-drilling have been made. Higher penetration rates in hard rock have been achieved by machines of lighter weight, higher piston speeds (up to 2,500 blows per minute), high rotation speeds, and, of course, better bit design. These machines, for underground use, operate on compressed air at pressures ranging from 100 to 150 pounds per square inch and are used for drilling blastholes of about 1.4 inches diameter up to 15 feet deep. Larger models, using coupled steel, drill 2.75-inch diameter holes up to 120 feet deep.

With these units, underground holes can be drilled at any angle. For vertical "down" holes, hand-held jackhammers are used; for vertical "up" holes, the machine is equipped with a telescopic in-line air feed leg; and for horizontal holes, the "drifter" machine is either mounted in a fixed cradle on a fixed tubular jack bar, or used with a hinge-mounted air feed leg, developed by Swedish manufacturers after World War II. For large headings, a battery of these machines is mounted on a mobile drill carriage termed a "jumbo." Such a jumbo is operated by one man with a console of hydraulic controls and with the capability of speedy manipulation in order to drill holes in the face in a wide range of angles and positions. Arising from this hydraulic system, a new family of percussive rockdrills is currently being developed to operate under hydraulic power instead of compressed air.

Naturally, these rockdrilling machines would not have been successfully developed unless the bit performance had also been improved. Early bits were formed on drill steel by forging and tempering. Some of these were chisel-bits and others were cross-bits. Until 1950, all were formed in special forging machines, operated by compressed air. Such bits had a limited life in hard rock, sometimes drilling only about 12 inches before being returned to the surface for resharpening. In the larger mines, many tons of drill steel were in transit up and down the shaft each day. In other cases, detachable alloy bits were used, but these were not successful in hard rock because of deformation of the joint.

With the advent of tungsten carbide alloy, following development in Germany during World War II for machining gun barrels, bit design was revolutionized. Drill steel is now available with bits formed from tungsten carbide alloy inserts. Drilling speed is considerably improved and bit life is very greatly increased.

Mobile surface blasthole drills have also experienced revolutionary developments in both the rotary and the percussive modes, more especially in "down-the-hole" drills.

Along with developments in blasthole drilling, there have been substantial developments in types of explosives since black powder was first used in underground mines in 1627. Nitrocellulose and nitroglycerine were developed by 1846, but these were not used for rock blasting until Alfred Nobel invented the fulminate detonator in 1867, as a safe, reliable method of initiating these high-powered explosives. In the same year, Nobel made liquid nitroglycerine safe to handle by absorbing it in kieselguhr (diatomaceous earth). The resulting explosive powder contained 75 percent of nitroglycerine. It was known as "guhr dynamite."

In 1875, Nobel produced blasting gelatine. This is a gelatinous mixture of 92 percent of nitroglycerine and 8 percent of nitrocellulose. Blasting gelatine has a very high water resistance. It is still one of the most powerful of the commercial explosives.

As the use of explosives for breaking rock developed, Nobel realized that there was a need for lower strength explosives. Therefore, in 1879 he replaced some of the nitroglycerine with sodium nitrate and other ingredients. A wide range of lower strength explosives has since been developed.

Nobel's old laboratory near Stockholm is still in effective use, although it has been rebuilt, extended, and modernized. Research work conducted by him and his followers in Sweden still leads the world. But during his career, Nobel had political problems. He was roundly criticized for producing such instruments of destruction and war. To counter these accusations, he instituted the Nobel Peace Prize.

As blasting procedures developed further, the pressure of competition and the need to reduce costs stimulated the search for a source of explosive energy cheaper than nitroglycerine. As far back as 1867 two Swedes (Ohlsson and Norrbin) took out a patent for an explosive mixture based on ammonium nitrate sensitized with various carbonaceous materials. Probably because it was not water-resistant, it did not become accepted until 1935 when it was packaged in sealed metal cans.

Then, in 1955, R. I. Akre developed the use of "Akremite." It consisted of a cheap grade of ammonium nitrate and carbon black. These experiments opened the way for the revolutionary development of a new range of explosive mixtures, based upon ammonium nitrate instead of nitroglycerine. One of the most popular mixtures is 94 percent of ammonium nitrate and 6 percent of fuel oil, known in the industry as ANFO. Where this new blasting agent is applicable (i.e., in charges free of standing water), it gives as good a performance as the dynamites at about one-third of the cost.

However, the next development was reflected in the need to promote the introduction of a blasting agent with adequate water resistance. This was achieved by Dr. Melvin A. Cook of the University of Utah. In 1960, he introduced the first of a family of ammonium nitrate slurries. He produced a saturated aqueous solution of ammonium nitrate and dispersed in this medium a mixture of excess ammonium nitrate and a carbonaceous "fuel" sensitizer, the whole being thickened by a gelling agent. A wide range of slurry blasting agents has since been developed, some of them metallized with aluminum powder to promote greater explosive energy output. So, whereas Nobel was recognized as the Father of Dynamite in the 1860s, we find a similar need to recognize Cook as the Father of Slurries one hundred years later.

Over 80 percent of the commercial explosives currently used in the United States are blasting agents: either ANFO or slurries.

A discussion on the history of improvements in blasthole drilling would not be complete without a parallel discussion on the development of exploration drilling. Until about 1900, most exploration drills were based upon the percussive churn drill, also known as the keystone or cable drill, for deep vertical holes, especially where hard rock was encountered.

Rotary drilling in hard rock, utilizing the scraping action of a drag bit, was not effective. But diamond drills invented by Rudolph Leschot of France in 1864 (for drilling the hard rock in the Mont Cenis tunnel) became very effective although costly. The cheaper Calyx drill introduced in 1904 was suitable in rock of medium hardness. But calyx drills were superseded by the tri-cone rotary roller bits introduced in 1909. The use of these bits has greatly improved rates and costs of drilling deep oil wells, and has considerably improved the feasibility of drilling hard formations for blasthole purposes. Many design improvements have since been made, including the design of the cutting teeth, the bearings, and the use of hard metal inserts where necessary. Assemblies of tri-cone rotary roller bits are now being used for full face drilling development headings in underground mines, in a variety of specialized mountings.

Meanwhile, diamond drills have been developed to greater levels of sophistication for core recovery purposes. Advances have been made in more accurate methods of surveying holes, improved core recovery, greater drilling speed, better hole alignment, and lower equivalent costs per foot, especially following the use of small sizes of industrial diamonds and ready-set bits. Drilling completion costs have also been improved immeasurably by the introduction of the wire-line principle of withdrawing the core, instead of intermittently raising and lowering the whole drill string whenever the core barrel is filled.

18 Ventilation

In the early days of underground mining, such as in Egyptian gold mines, the circulation of ventilating air was very poor, mainly because of the small dimensions of the passageways. Air circulation was later improved, especially at Mitterberg, when an additional shaft was sunk to provide a separate exit for the foul air.

In some cases, as at Laurium, duplicate passageways were run in parallel and connected by short crosscuts at intervals to improve the circulation of air, which up to this time depended upon the natural ventilating pressure. The practice of fire-setting was a significant factor in vitiating the mine atmosphere.

Agricola describes the earliest methods used for forced air circulation. One method involved the use of a series of bellows, operated manually, to force a current of air through the workings. Another device was a drum, rotated by a water wheel. An opening at the periphery of the drum caused air to flow along a series of ducts. Perhaps this was the forerunner of the centrifugal fan.

During the 19th century, with the increased activity in coal mining, ventilation problems became acute, especially in the deeper workings. Two main problems were noted. One was the prevalence of chokedamp: a mixture of nitrogen and carbon dioxide, resulting in a reduction in oxygen content. This caused asphyxiation. Another was the presence of firedamp: a mixture of methane and air, which could be readily exploded by a naked flame when the mixture contained between 5 and 15 percent of methane.

To overcome these problems, an additional ventilating shaft was sunk so that fresh air entering the main shaft could be exhausted up the upcast ventilating shaft, without being short-circuited. Nevertheless, the quantity of air in circulation was in many cases still too low. This flow was increased by using a brazier fire, or furnace, at the collar of the upcast shaft to induce more air to flow, a method used in the Belgian coalfields prior to 1650. The Belgian system enclosed the brazier within a chimney about 5 feet square and 30 feet high to increase the draught. Where there was little or no methane, the fire was maintained at the bottom of the upcast shaft so that the shaft itself could act as a chimney.

As mine workings extended beyond the influence of the two shafts, it was found that many of the coal faces were inadequately ventilated. To induce the main air current to flow around these peripheral workings, a series of air doors was built to direct and control the flow. This system was first introduced in Cumberland, England.

The next step, following the development of machines during the Industrial Revolution, was to introduce mechanical ventilators to pump fresh air into the mines, or to exhaust foul air. A primitive auxiliary fan called a "Blow George" and operated by three men was first introduced to ventilate blind headings. Later models were belt-driven by steam engines. The first air pump for ventilation purposes was introduced by Buddle in 1803. It could handle up to 600 cubic feet of air per minute.

Belgian and French engineers were the first to develop main ventilators on the continent. The first fan (air pump) was installed at the St. Louis Mine in the Mons district in 1830. It inspired the Belgian Academy of Science in 1840 to offer prizes for the development of improved models. This effort was so successful that furnace ventilation was abandoned by 1850.

In England in 1835 an exhaust fan was invented by William Fourness of Leeds. By 1845, it had been widely adopted. More elaborate and larger fans were developed in following years.

The problem of coal mine explosions had become especially acute during the 18th century. Ventilating fans at that time were not always of sufficient capacity to disperse large accumulations of methane gas. During this time, the practice of "firing" small gas flows had been adopted. A "fireman," covered with a wet cloth, crawled near the face and poked a candle attached to the end of a long pole into a blower of gas. This, of course, was a dangerous practice.

It had been realized that many explosions were triggered by naked lights used for illuminating purposes. Many of these explosions resulted in multiple fatalities. In order to reduce this hazard, alternative sources of illumination were developed (see chapter 19).

Yet even the introduction of the safety lamp did not prevent all gas explosions, because some miners persisted in the use of candles, and because lamps were used improperly.

To reduce the incidence of gas explosions triggered by explosives, a special type of "permitted" explosive was introduced to coal mines in Britain in 1896.

By 1890, it was realized that methane gas was not the only cause of mine explosions. Coal dust was now recognized as another cause, either alone with air or when mixed with methane. Except for a few that are exceedingly wet, coal mines are dusty. The various cutting, firing, loading, and haulage operations all create dust, and this is carried through the mine by the flow of ventilating air. With sufficient fine coal dust in the air, a hot flame will ignite it and create an explosion.

Several serious explosions occurred in coal mines in England and France in which methane was not present. The very great hazard of coal dust was slowly recognized. In 1911, "stone dusting" was introduced in British coal mines to smother accumulations of fine dust. In Honkeiko, China, more than 1,000 lives were lost in 1941 because of a coal dust explosion. Coal dust was one of the factors associated with the loss of 439 lives in a mine disaster in Glamorgan, Wales, in 1913.

The first reported coal mine explosion in the United States was in Virginia in 1810. Until 1890, there were 851 fatalities due to 43 explosions in ten states. Between 1891 and 1900, 1,006 fatalities were reported from 38 explosions and 426 from 250 minor occurrences.

A number of disastrous coal mine explosions (from either cause) occurred in eight coal mines in the United States in 1907-08, resulting collectively in the death of 1,148 miners. One such explosion occurred in a mine near Monongah, West Virginia, in December 1907, when 362 lives were lost. As a result, the U.S. Bureau of Mines was established by an act of Congress and became effective in 1910.

Since 1950, tremendous advances have been made in ventilation technology. It is now realized that miners can work at maximum efficiency only under comfortable conditions. In the deep gold mines of South Africa, where rock temperatures of 140 degrees Fahrenheit are not unusual, the ventilating air is air-conditioned, with huge refrigerating plants established in strategic locations underground.

19 Mine Illumination

In the early days of underground gold mining in Egypt, it was common practice to use oil lamps for illumination. These consisted of open earthenware vessels containing animal or vegetable oil. Figure 19.1 shows two styles of miner's lamps used during the Roman era. That on the left is made of brass; an earthenware (white clay) lamp is shown on the right. Later, a flame-wick was added to open lamps of this type. When lit, the resulting flame gave forth a certain amount of luminescence. In some cases, a smaller version was fastened to the miner's forehead. Torches made from bulrushes soaked in oil were also used for general illumination.

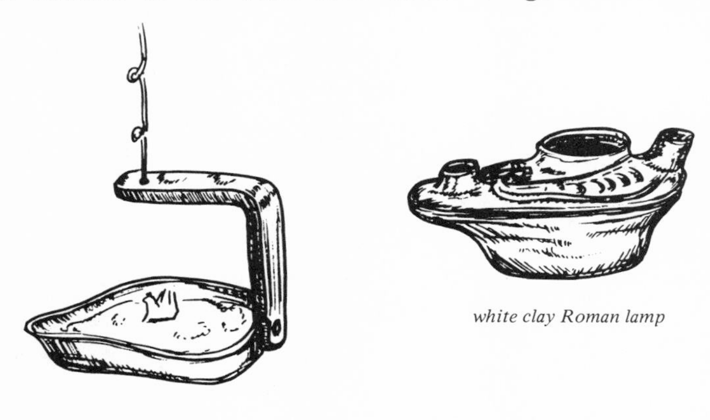

Fig. 19.1. Two types of Roman lamps.

Miners in the 16th century in Europe used to take a bundle of slivers of a resinous variety of pine wood (spruce or fir) down the mine to provide lighting during the shift. By lighting one end of a pine chip, it burned slowly and emitted a certain amount of light. The pine chip, about 18 inches long at the beginning, was usually held between the teeth to allow the use of both hands for work.

After it was found that a lamp with an open flame was liable to trigger a disastrous explosion in coal mine atmospheres carrying significant methane gas, alternative methods became necessary. The general aim was to produce a lamp with a cooler flame than the threshold ignition temperature of methane. This was achieved in two different ways.

1. The faint phosphorescent glow derived from putrescent fish skins or decaying fish bones was used.
2. In 1760, Carlisle Spedding invented the steel mill. This was the first mechanical apparatus for producing light. The shower of sparks emitted when a steel wheel was cranked by hand against a piece of flint gave forth a certain amount of light, but was not supposed to generate sufficient heat to ignite methane (see figure 19.2).

Fig. 19.2. A steel mill.

Both of these illuminating methods offered a substandard intensity of illumination, however, and no doubt contributed to the incidence of the occupational disease known as nystagmus, evidenced by a rapid oscillation of the eyeballs, perhaps a natural reaction of the eye mechanism seeking to seize every flash of light available.

The Spedding steel mill was regarded as the better alternative but later it was found that explosions still occurred; these were due to the presence of fine coal dust in a dry atmosphere rather than to the presence of methane.

The next step was the development of the "safety lamp." In 1798, Humboldt introduced his lamp but it would not burn steadily in impure air. Trevithick produced a later model in 1811. In 1813, Dr. William Reid Clanny produced a safety lamp. It consisted of a candle in a closed metal case fitted with a semicircular glass and was fed with air provided by a bellows. The Clanny lamp had certain disadvantages and these were overcome by a later model produced by Sir Humphry Davy in 1815. In the Davy lamp, the candle was replaced by an oil lamp enclosed in a gauze cylinder to disperse the heat to a temperature below the flash point of methane.

The Davy safety lamp is regarded as the first practicable lamp that could be safely used in gassy coal mines. It was further regarded and used as an instrument for detecting and measuring the concentration of methane at any point in a coal mine. Subsequent improvements have been made by George Stephenson (1815, 1817, and later), J. Meuseler (1840), F. Eloin (1850), J. B. Marsaut (1871), A. Boty (1881), C. H. Wolf (1883), and many others.

Nevertheless, in other than gassy mines, the oil lamp gave way to tallow burners, and then to candles, which were made of tallow and provided with a better wick for more efficient burning. The candlemaker on each mine was known as a chandler. Candles could be fixed onto the rock wall of a working face or onto adjacent timber by means of a metal holder provided with a hook and a spike. This was known as a spider (see figure 19.3). For more portable illumination, a candle was mounted on the brim of the miner's hat and kept in position by a daub of clay. But as the speed of air flowing in mine airways increased to provide better ventilation, candles were difficult to keep lit. By about 1900, most metal mines used acetylene lamps. These consisted of a lower chamber containing calcium carbide and an upper vessel containing water (see figure 12.5). Water was allowed to drip through an adjustable needle valve onto the carbide, thereby generating acetylene gas which passed through a small tube to an external gas burner. Such lamps, especially when the burner was backed by a polished reflector, gave a great deal more light than a candle, and were much more practicable. They were provided with a handle and a hook, for carrying or suspen-

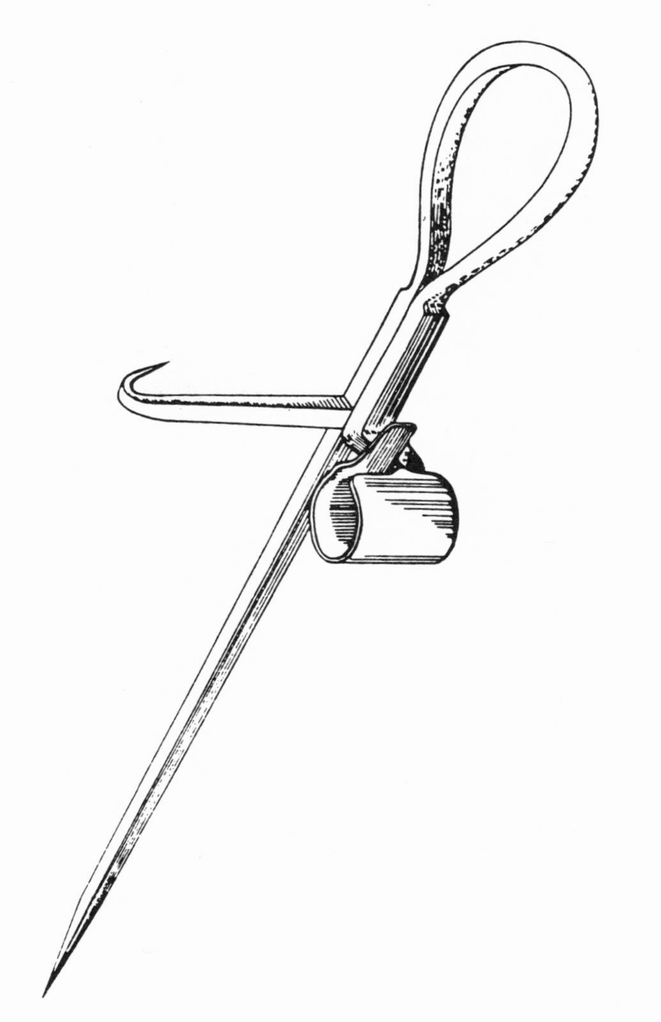

Fig. 19.3. A miner's spider (classic miner's candleholder).

sion. A smaller unit could be mounted on the brim of a hat. Lamps were passed in to the lamp maintenance shop on the surface at the end of each shift, and serviced and recharged for the corresponding shift on the following day.

From 1920 onward, attempts were made to produce an even greater intensity of illumination because it was realized that fewer accidents would ensue. An electric cap lamp, clipped to a miner's helmet, is now connected with a short cable to a battery mounted on the miner's belt. Batteries are recharged on the surface at the end of each shift. These lamps are now used for both coal and metal mines, and the safety lamp is used only for detecting and measuring the methane content of the air, although a battery-operated instrument known as a methanometer is now in general use for this purpose.

Meanwhile, in all main work areas of a mine, permanent electric lighting is now installed, using either gas-filled incandescent lamps or fluorescent strip lights. In these areas, the rock walls are painted white in order to gain more reflected light. In many work faces where mechanized equipment is employed, electric floodlights are arranged to give increased illumination.

20 Mining Law

The history of ownership of mineral deposits has evolved in different patterns over past centuries. At first, mineral rights were associated with the ownership of the surface land below which they occurred. Many early tribes held the land in the common interest of its members. At a later stage, as tribal leaders or rulers emerged, the land was transferred to them in the communal interests of the tribe. Egyptian mines became the property of the pharaoh. This was the beginning of the regalian system in which mineral rights were vested in the ruler.

In Greece, the minerals were regarded as state property for the benefit of the citizens, irrespective of land ownership. But the mines were not operated by the state. They were leased to particular persons or companies for up to ten years by a board of magistrates; the leases were administered under a mining code by a director of mines. Leaseholders were required to make an initial payment and pay a rent of 1/24 of the value of bullion produced. At Laurium, the silver-lead mines became the economic mainstay of Athens, which thereby flourished as a trading center. Laws were enacted to ensure that lessees operated the deposits in the best interests of the state.

In Roman times, the question of mineral rights was bound into the ownership of the surface land. This was called the accession system. Nevertheless, mining was discouraged on the Italian mainland although saltworks were established and leased. Any gold produced was the property of the state. In the colonies conquered by the Roman legions, the mineral ownership was vested in the state and generally leased to Roman citizens upon payment of a rent, or royalty. Many leaseholders made large fortunes. This led to a system of bidding for leases. During the 6th century, the usufruct system was introduced. It granted mining rights on other people's land provided no surface damage was done. As the Roman Empire began to decline, and the need for precious metals became urgent to avoid currency devaluation, the usufruct system was abandoned and state officials were appointed to operate the mines, with substantial incentives to expand production. Following the collapse of the Roman Empire and the advent of the Dark Ages, the feudal system developed.

In England, the owner of the surface land also owned the mineral rights, except where common law recognized the rights of miners to operate on private land, as in Cornwall (under the provision of the Stannary Courts), or in Derbyshire and a few other places. A royalty was payable to the owner. However, the king owned all the gold and silver deposits, presumably for raw material for coinage to feed the state treasury. But this did not apply to gold and silver recovered from base metal mines. The practice was not uniform throughout the mining districts of England, and from time to time amendments and adjustments were made.

In Germany, mining districts were controlled by a Bergmeister in the name of the ruler, who possessed both the land and the mineral rights. Claims were staked out on the ground and the area was granted under lease for mining if the bergmeister so approved. For initial discoveries, a larger area was allowed. The miner (grantee) had the right to follow the vein downward even if it passed beyond the vertical extension of the claim boundaries. The miner paid a tithe, or royalty, to the ruling prince.

Following the Middle Ages, two distinct systems evolved from the Roman regalian theory, a theory based upon the right of the state to be paid a royalty, since all deposits of precious metals were deemed to be the property of the Crown.

1. On the European continent, the ruler owned all the mineral rights, and the surface landowner was merely to be compensated for damages.
2. In England, the landowner also owned the mineral rights; but if he were unwilling to work the mineral deposits, others could be granted this privilege.

Following the Australian gold rush in 1851, the right of the Crown to the precious metals was hotly contested by the miners. When the government forced them to purchase a license to mine gold, they rebelled.

In Spain, the right of the sovereign to precious metal deposits and to a 20 percent royalty (after 1504) was readily accepted as a source of revenue for coinage, defense, and general government funds to offset the need for taxes. But in 1783, Charles III decreed that all mines were the property of the Crown, thereby separating land ownership from mineral rights. In such a case, a 20 percent royalty was due to the Crown for all type of minerals. In Mexico and most Latin American countries, the law generally follows Spanish practice. In France, the regalian system was modified differently. There was to be no question of mineral ownership until an act of discovery occurred. Although mineral deposits were not owned by the state, they were so controlled. When a mineral deposit was discovered, the state awarded a concession. If the discoverer was not the landowner, then the latter was to be compensated by way of royalty.

Mining law in Canada is based upon the principle that each province has jurisdiction over mineral property within its borders. In all provinces except Quebec, the landowner retains the mineral rights except for gold and silver. In Quebec, the mineral rights are not necessarily conveyed with the land unless specifically provided. Some provinces grant patent rights to mineral claims and others merely lease the rights.

In the United States, the law governing mineral rights is exceedingly complex. Under colonial law in the original 13 colonies, the mineral rights were associated with the land charter, and royalties, generally of the order of 20 percent, were payable to the Crown for gold and silver, or to the charter holder for other minerals. After 1776, the lands not already privately owned or sold by the government were reserved as the public domain. The area of the public domain grew significantly because of various acquisition treaties and purchases, ranging from the Louisiana Purchase in 1803 to the purchase of Alaska in 1867. Otherwise, some territories such as Hawaii (formerly), Puerto Rico, Guam, and the Virgin Islands did not become part of the public domain, nor did public lands in Texas which entered the Union in 1845.

In the course of time, the area of lands included in the public domain became progressively reduced by the federal government in the shape of grants for school lands, railroad lands, swamp lands, and lands alienated under the Homestead Law of 1862. Furthermore, severe restrictions were placed on prospecting or mining on other parts of the public domain, such as on Indian reservations, military reservations, national parks, national forests, power and water supply reservations, town sites, federally acquired lands, and lands acquired under the Wilderness Act of 1964.

Apart from the above restrictions, the public domain, under the Mineral Location Law of 1872, is open for citizens to enter upon the land, prospect for minerals, make discoveries, stake claims of a certain size, record the claims, work the deposit, and sell the mineral products. No specific title is granted for a claim unless and until it is patented. This includes the surface rights. Distinctions are made between lode and placer claims. With lode claims, the miner has the right to work the lode in depth even if it dips into an adjoining claim beyond the vertical projection of the side lines of his own claim. This provision was adopted from early German practice. Subsequent litigation over contested rights has resulted in a tremendous waste of mining corporation finances. This principle of extralateral rights represents an abominable feature of American mining law, but it has not been excised. There are other serious deficiencies in the Act of 1872, as recognized by the Hoover Commission in 1949, but none have been removed.

Meanwhile, special provisions have been made for those mineral substances that cannot readily be classified as lode or placer deposits. Accordingly, legislation has been enacted for patent procedures on coal deposits (1873), building stone (1892), oil deposits (1897), saline materials (1901), and common varieties (of minerals such as stone, sand, and gravel, in 1955). Some of this legislation was absorbed by the Mineral Leasing Act of 1920.

When certain public lands, other than the public domain, are leased, a royalty is payable. Similarly, for lease agreements involving mining on private lands, a royalty payment is negotiated with the landowner. In some states where the surface rights have been separated from the mineral rights, a severance tax is applied. For ordinary claims on the public domain, no royalty is payable.

This information on American mining law is necessarily superficial. For more complete details, the reader is referred to American Law of Mining (5 volumes) by Bender.

21 Working Conditions

Wide variations in working conditions and in the social status of miners prevailed throughout the millennia of mining history. During the earliest days of organized mining, as in the Egyptian goldfields, the miners were prisoners of war, criminals, slaves, and political reactionaries. Many of the latter were sent to the gold mines. Sometimes their families accompanied them, perhaps for similar employment opportunities.

A Sicilian historian, Diodorus, has described some of the conditions. He stated that the miners had to work constantly throughout the 24 hours of the day. They were guarded carefully to prevent their escape, usually by foreign soldiers with whom there was little possibility of communication because of the language barrier. The work was described as arduous, exhausting, and uncomfortable in terms of the limited work space and the resulting bodily postures. No rest periods were allowed even due to sickness or infirmity.

At Laurium, in the Greek silver-lead mines, slaves were also employed. Thucydides writes that following the invasion of the Spartans in 413 B.C., 20,000 slaves escaped. Many of them were employed in the mines. During the Roman occupation of Spain in the 2nd century A.D., the mines at Murcia employed over 40,000 men. Most of these workers were believed to be slaves. Slaves were used as motive power for operating the water wheels, in treadmill fashion, and for other manual tasks.

Yet the development of technical skills was sponsored by the Romans. They trained mining engineers at the Rio Tinto mines for supervisory positions throughout the Empire.

Still, the status of miners up to the 5th century A.D. was very inferior. The majority of miners were slaves. Some lived underground. An underground mine was a useful incarceration compound because avenues of escape were extremely limited. Some of the Roman miners lived in caves near the mine entrances. Evidence of chains and fetters have been found - also chained skeletons. The death rate was high. Slaves lived and worked under atrocious conditions. Even for a slave, mining was considered as the lowest form of employment.

During the Middle Ages, the feudal manorial system developed. Here the serfs were bonded to the landlord primarily for protection against brigands and marauding tribes. By the 12th century, the bishops and landlords gradually freed the serfs to search for mineral deposits, which they came to regard as a major source of income. In this way, individual serfs became prospectors and miners and many new mineral deposits were found, especially in the Erzgebirge. This occurred during, and was mainly responsible for, the mining revival from the 11th to the 14th centuries.

When political conditions stabilized after the Hundred Years' War (1337-1453) and the Hussite Wars (1415-1435), renewed mining activity brought the miner into his own, with substantial social privileges. German miners had become free men and were exempt from taxation and from military service. These special privileges were granted because of the importance of the skills they possessed, and because mineral production was recognized to be of supreme national importance.

In England, the tin miners of Cornwall were given a charter in 1201 by King John to prospect on any land (except orchards, graveyards, churchyards, and highways) on payment of a tribute of 10 percent of their revenue to the landowner. This charter was reissued by Edward I in 1305. These tin miners had their own courts of law, under the sole jurisdiction of a "mining warden," a sort of magistrate or judge. These courts were known as the Stannary Courts. Tin miners could not be compelled to appear in any other court. They also had their own councils, with the right to legislate in their own special interests. Many of these privileges were developed as far back as 950 A.D. The lead miners of Derbyshire, Alston Moor, and the Mendip Hills also had their own courts.

In France, King Philip IV (1285-1314) freed the serfs and thereby ended all forced labor in French mines. In Germany, many mining towns, such as Freiberg, were granted the status of "free cities." This meant that the whole community of miners had freedom of occupation, freedom of worship, and special privileges such as freedom from liability for taxes and military service. Miners represented the upper class, just below the nobility. They could bake, brew, and butcher as they wished, with complete freedom of trade. They were also free to work in any mine in any district of their choice.

As early as 1518, miners had free accident insurance, with continuing pay up to eight weeks of incapacity. There were also welfare plans for the aged and the disabled, as well as for those who developed lung disease. Funeral benefits, sick pay, and life insurance were granted to miners and their families. Many owned their homes. These privileges were granted originally by associations of free miners who clubbed together to operate a particular mining property.

Miners were paid even when the mines for any reason were temporarily closed. Extra pay was awarded for overtime, for working on holidays, or for working in wet places. They were also paid for 34 public holidays (feast days) a year, even if they did no work. Shifts were normally of seven hours.

After the 16th century, however, the scale of mining operations grew enormously because large amounts of capital had become necessary to develop and operate the mines at deeper levels. Mining under these conditions had progressed beyond the scope of a self-employed individual miner, or even a group of miners. Mining companies were now being formed to operate mines, each by a group of 128 shareholders. These were mostly businessmen (absentee stockholders) who contributed the capital and of course collected the profits.

Actual mining operations were then controlled by paid managers and foremen who hired the miners and other workmen. In this way a clear distinction emerged between the mine owners and the miners. This meant a gradual loss of privileges. This loss of status was most marked in central Europe. It often led to bitter disputes between workers and management.

In the 17th century, a similar change in status occurred in England. Many of the independent tinners of Cornwall and the lead miners of Derbyshire had to relinquish their own mines to capitalists, and had to become wage laborers in their employ.

Nevertheless, the miner retained a recognized place in society. Agricola amply justifies mining and underlines the merits of the mining profession. In certain instances, wage-earning miners were treated with consideration. One such humanitarian employer was the London Lead Company which was formed during the reign of Charles II. This company became the largest and most technically progressive producer of lead and silver in Europe. It owned various mining properties in Wales and Derbyshire. This was an enlightened employer with the well-being of its workmen and their families at heart. In a spirit of altruism and enlightened self-interest, it maintained a well-fed, well-housed, healthy labor force in a high state of morale, fostering promotion from the ranks and a general spirit of independence. On several occasions it subsidized food supplies. Following the depression of 1815, the company continued working at a loss in order to avoid retrenchment of its employees.

The London Lead Company also provided amenities such as reading rooms, libraries, bunkhouses, and cottages; and improved the water supply by providing piped water. In addition, it furnished free medical care, a social security fund, and old age pensions.

Another great mining company, the Beaumont Company, which operated 50 mines, displayed a deep concern for the welfare of its employees. From about 1700 to 1882, both these companies had control of lead mining in the north of England.

Nevertheless, the employment of women and children underground during this period provided room for serious reforms. In 1813, the employment of children under 10 years of age was prohibited in French and Belgian mines. In 1839 a similar enactment was made in Germany for children under 9, and extended to children under the age of 12 years in 1882. An act was passed in Britain in 1842 forbidding the employment underground of women and of children under 13 years of age. Following the Civil War in the United States and the abolition of slavery, children were employed underground in Pennsylvania coal mines.

However, opportunities for an independent miner to make his fortune by his own individual efforts occurred again in the mid-19th century. From 1848, the discoveries of rich gold placer deposits in California and neighboring western states, in British Columbia, in Australia, and in New Guinea, brought men from near and far to try their luck. No great amount of experience was required. Similar opportunities arose with diamond deposits in South Africa and stream tin deposits in Malaysia.

But generally, all these deposits in time became exhausted; or else subsequent deep vein mining led to operation by mining companies and most miners became wage-paid employees. Various periods of violent bargaining led to the formation of labor unions. Many violent strikes in most countries have led to much bitterness between employer and employee. But by and large, the miner, through his union, has been able to receive a reasonable pay scale and an enhanced package of fringe benefits.

IV

Traditions, Customs, and Folklore of Miners

An Introduction to Part IV

The ageless spirit and philosophy of mining has set the stage for the establishment and continuance of many particular customs and rich traditions, derived from the earliest days of mining, but more especially from the 14th to the 16th centuries in Europe.

In his magnificent encyclopedic volume dealing with the spirit and history of mining (Mining Lore, 1970), Dr. Wolfgang Paul underlines the little-realized fact that mining, together with its traditions and cultural background, has been serving mankind in the production of fuels and other basic mineral materials for many thousands of years, and is even more important to society today than it ever was.

Mining activities form the core of man's historical development; they also provide the means by which the standard of living of any country can advance and be sustained at a level substantially above that of a purely agricultural (peasant) economy. Unfortunately, this fact is neither realized nor appreciated by the present-day comfort-prone citizenry of most industrialized countries. Without mineral production, their smugly taken-for-granted comfort levels and ultramodern conveniences would have been impossible to attain.

Across all languages and international boundaries there has developed a sort of freemasonry among miners. The esprit de corps and atmosphere associated with mining activity is significant. Ten thousand years of high-ranking, high-order community service in the development and maintenance of civilization is not easily eroded. Its spirit and traditions have kept it alive.

We will discuss a few of these noteworthy traditions here. One such custom that can be readily maintained with a delicate spirit of dedication is to pay homage regularly to St. Barbara as the patron saint of miners.

22 The Legend of St. Barbara

Barbara lived in the 3rd century A.D. in Nicomedia, the capital of the Roman province of Bithynia (Asia Minor). She was the only child of Dioscuros, a high-ranking and wealthy man. Her father adored her, had her tutored in the best schools of arts and science, and set out to reinforce her faith in the Roman-Greek gods.

To protect Barbara from foreign influences, he provided sumptuous living quarters for her in a tower. But her extreme loneliness caused Barbara to think seriously; as a result, she became more and more convinced that the old gods were but hollow imitations.

Without her father's knowledge, she became familiar with the teachings of Christ and had herself baptized. At that time, Christians were being persecuted nearly everywhere, and considered as enemies of the state. Adherence to Christianity was subject to the severest punishment. (An example of these repressive measures is given by Pliny the Younger. When serving as the Roman governor of Bithynia he reported to Emperor Trajan that a group of people calling themselves Christians was gaining strength in the province. "The method I have observed towards Christians is this. I asked them whether they were Christians. If they admitted it, I repeated the question twice and threatened them with punishment. If they persisted, I ordered them to be at once punished.")

Dioscuros planned to marry Barbara to a very prosperous man with a view to increasing the family fortunes. At first Barbara asked for time to reflect. Following his return from a long journey, Barbara explained to her father that she was now a Christian and did not wish to marry. She had already removed the different images of pagan gods from her penthouse and had replaced them with crucifixes.

Dioscuros, seeing that his only child had turned to the hated new religion and that he himself had been placed at a disadvantage, was overcome with rage. He handed over his daughter for the assessment of punishment to the Roman proconsul Martianus, a supreme court judge.

Martianus tried at first by kind persuasion to make her break with her faith; but when this failed, he had her thrashed and cast into jail. Due to the strength of her faith, her wounds healed immediately.

On the following day, she was ordered by Martianus to pay sacrifice to the pagan gods. When she refused, she was mutilated in a dreadful way. When she continued to proclaim her Christian faith, she was sentenced to die by the sword.

Barbara went to her place of execution in cheerful ecstasy in her enthusiasm for her true faith. Her last wish was that God, through her experience, might help all those confronted with and unprepared for a sudden, untimely death.

Barbara's barbarous father was so outraged that he himself severed his daughter's head! Immediately following Barbara's death, a terrible thunderstorm arose. As punishment for his monstrous crime, Dioscuros was killed by lightning.

This is the way the legend goes. In this form, it represents the oldest source of information.

According to a later version (late Middle Ages), the story took place near Athens. Barbara was supposed to have escaped from her prison, a tower, and to have found refuge with the miners of Laurium who protected her in the mine. But when she climbed up the shaft, she was intercepted by her persecutors and decapitated by her own father.

The precise year of St. Barbara's death is unknown. Some assume the year was 304 or 306 A.D. Other sources give an earlier year (254). It is known that from the year 303 on, persecutions of Christians did take place in the province of Bithynia. Nicomedia is situated at or near Izmit, at the head of a gulf in the Sea of Marmara (Turkey).

PROTECTRESS OF MINERS

Later, when Christianity had become firmly established, St. Barbara was invoked as a protectress against the perils of lightning. The belief had become widespread that Barbara could control thunder and lightning and other manifestations of noisy flame and fire. A custom developed of ringing the church bells during thunderstorms, in order to remind the faithful to pray in her name.

Due to her reputed power of remission of sins, Barbara also became the protectress against sudden unrepentant death, as may be experienced by miners. For centuries past, there had been a spiritual need of mankind, especially of those working underground, for help to cope with the unknown rulings of a higher destiny. They, therefore, sought an intercessor, a patron who might shield them from day-to-day problems and the perils of life. Miners formed a large part of those for whom Barbara prayed in the hour of her death; and because the mining fraternity had to cope with many hazards to life in those days, Barbara was adopted as their patron saint by the miners of central Europe.

Miners later developed the use of gunpowder for disintegrating rock. This involved manifestations similar to thunder and lightning flashes. Miners, therefore, needed special protection against accidents arising from the use of explosives. The reputation of St. Barbara as their adopted patron saint was thereby strengthened.

THE ADORATION OF ST. BARBARA

In Western Europe, the Barbara veneration began in the Netherlands, probably in the 14th century. It spread to Germany, Bohemia, Hungary, and Italy. Barbara Day was listed as a holiday in the very earliest festival calendar of the city of Cologne.

In eastern Europe, the first reports arose in the 5th century. The first Barbara church was built in about 900 A.D. in Constantinople (now Istanbul, Turkey). But at the present time, it is impossible to determine precisely where the custom of the Barbara adoration first became entrenched. Possible sources are the mining districts of Freiberg (Saxony), Kuttenberg (Bohemia), and Schemnitz (Slovakia). The great Barbara Cathedral, built in the old Silver City between 1388 and 1518 and consecrated in 1483, supports this contention. The cathedral was built around an already existing Barbara Chapel. The abundance of old statues of Barbara (still in existence) in the former ore mining district of Schemnitz supports the third locale. According to an entry in 1346 in the register of the bishopric of Meissen, there were already 60 Barbara altars in Freiberg and the neighboring district. The cathedral in Freiberg had two Barbara altars, one probably dedicated by the miners' union. Two more Barbara altars were in the Peter's Church and in a fourth old church. In 1490, a mine owner built and dedicated a Barbara Chapel outside the Erbisch Gate.

The mining town of Kuttenberg for centuries has had, as its coat of arms, St. Barbara above the crossed hammer and gad, the classical symbol of mining. Numerous mines in many mining districts, such as in Freiberg, Saxony, and in other parts of the Erzgebirge, bore the name of St. Barbara. Names of mines frequently indicated wishes and hopes. In both Freiberg and Marienburg, there was a mine with the name of "St. Barbara Bonanza"; another one existed near Schneeberg. In Joachimsthal (now Jachymov), about a dozen lodes had the name of Barbara, among them a "St. Barbara High-grade" and a "St. Barbara Good Hope Vein." In the Harz Mountains and the Austrian alpine provinces there are also Barbara names for well-known veins, mines, portals, and churches.

THE FESTIVAL OF ST. BARBARA

In all European countries endowed with mineral wealth, and especially in those regions or provinces whose prosperity depends so much upon mineral production, the Festival of St. Barbara, the Patron Saint of Miners, is celebrated. Since the 14th century, St. Barbara Day has been celebrated every December 4, or on the next following Sunday, in the spirit of a mining festival. Only under extraordinary circumstances was such a celebration ever deferred. In some mining centers, even today, the traditional church service or Barbara benediction is still observed, followed in every case by a cheerful gathering of miners to drink ale, as one of their worthy customs.

Hundreds of thousands of miners in Austria, Germany, Italy, and parts of Poland, Czechoslovakia, Hungary, and other countries annually celebrate the Barbara Festival, whatever their religious affiliation. This serves to illustrate their enduring attachment to the Barbara faith as a deeply rooted mining tradition. It further indicates how firmly the mining profession is anchored in healthy traditions, based upon a spirit of professional pride.

REPRESENTATIONS OF ST. BARBARA IN THE ARTS

The wide dissemination of the Barbara tradition throughout the European mining industry would seem to justify a short description of the portrayal of St. Barbara in the arts. St. Barbara has a prominent role in the pictorial arts because some of the greatest masters sought to portray her as a model for their artistic endeavors. The most famous works stemmed from Dutch, German, and Italian schools. These included paintings by Palma Vecchio and Hans Holbein the Elder. Worthy of mention also are the works of several outstanding Austrian masters: artists whose names are still unknown.

Representations of St. Barbara are known to us in the form of several woodcuts (1415-1550), copper engravings (16th century), and brush drawings; there are also many paintings and sculptures, the most famous of which are from the 15th and 16th centuries. The oldest Barbara depiction is a Gothic glass window from the St. Leonhard Church in the Lavant Valley, Carinthia, Austria, from around 1370. In radiant, gorgeous colors, an everlasting masterpiece was here created by an artist unknown today. One can say that around 200 Barbara depictions still exist, forming a wide arc from the Gothic, through the Renaissance, into the Baroque. Roman figures are unknown. Scenes with mining motifs, for example with St. Barbara floating on clouds with miners and hoists and ropes below, are sometimes seen.

The Austrian city of Leoben, rich in mining lore, is a treasure chest of Barbara memorabilia. There are many others in Europe. In the City Square (Hauptplatz) at Leoben is a 17th-century fountain. Its centerpiece is a statue of St. Barbara adorned in miner's garb, wielding a miner's hammer. For many years it has become a tradition (almost a ritual) for seniors in mining engineering from the Montanuniversitaet Leoben, on graduation day, to leap across the font, one by one, climb the statue, and kiss St. Barbara, in gratitude for the completion of their professional qualifications. It is reported that many students, in vino on that eventful day, suddenly find their leaping capacity limited. The resulting accidental "splashdown" and wading procedures are enjoyed by citizens and students alike. It is an important annual event. But no student ever fails to complete his mission. Such is the strength of tradition!

The Bergakademie Freiberg in Saxony (East Germany) is the oldest mining school still in existence. It was once recognized as the foremost scientific academy in the world. Among its members were many celebrated scholars in the physical sciences. The city of Freiberg possesses many artistic representations of St. Barbara among its cultural attributes.

It is reassuring to note that the Canadian Institute of Mining and Metallurgy recognizes and honors St. Barbara by the annual award of the Sancta Barbara Medal to the wife or daughter of one of its members who has played a significant role in the progressive development of Canadian mining communities.

In the United States, a disastrous explosion occurred December 6, 1907, in West Virginia at the Monongah Coal Mine. As a result, 362 miners lost their lives. The last ton of coal won from the mine was set up as a monument, and the St. Barbara Nursing Home was established as a fitting memorial.

Otherwise, the soul of St. Barbara seems to remain unheralded outside of Europe. Nevertheless, the existing array of St. Barbara memorabilia and the associated European customs and culture provide a well-established foundation upon which the mining profession in younger countries can readily build. What can be done to preserve and support the commemoration of this beautiful Barbara tradition, in these modern times when the mining profession needs more and greater reliance on its spiritual traditions?

23 Other Mining Traditions

RELIGIOUS TRADITIONS

There are many old churches in central Europe reflecting the religious interests and instincts of miners. Some churches in the Saxon-Bohemian Erzgebirge feature altars, pulpits, statues, figures, and stained-glass windows representing miners in their characteristic garb and with their tools. These have been executed in wood, stone, metal, and glass, and many are gilded.

Even outside the church religion prevailed. Each mine had its simple "praying room" in which miners, going below to work, regularly prayed for an accident-free shift; and on returning up the shaft at the end of the shift they offered their thanks to the Lord. In the Valier Coal Mine, near Du Quoin, Illinois, in 1923, it was the custom for the miners to gather at the bottom of the shaft, before dispersing to their work places, to hold a prayer service and sing a hymn. In Cornwall, the miners were regular attenders at Sunday Chapel and the Welsh coal miners sang their hymns with great devotion.

Three churches are associated with mining activities. In the salt mines of Wieliczka, Poland, established in the 11th century and still working, is an underground church on seven different levels from two shafts with direct connections to the city center. The levels of the mine consist of a labyrinth of chambers and passages, totalling 65 miles, and are inter-connected with flights of steps. Many of the old chambers, hewn out of salt, are embellished with portals, statues, and candelabra. There are also two large chapels containing altars and ornaments. In addition, a hospital deep in the mine treats patients for asthma and other allergies. The mines employ over 1,000 men and produce over 60,000 tons of salt per year.

A similar underground cathedral has been hewn out of salt in Colombia, near Bogota, in the Zipaquira salt mines. It is reported that the cathedral ceiling is nearly 100 feet above the floor. In 1956, a postage stamp was issued featuring this unusual cathedral.

In a coal mine 750 feet underground near Swansea, in South Wales, there is a church built by the miners from mine timbers and pit props, illuminated with miners' candles. Every Monday morning the miners gather for prayers, Bible readings, and hymns sung in the Welsh language.

MUSICAL TRADITIONS

Miners also developed a great attachment to music, and many beautiful hymns and songs have been written in testimony of their work for the benefit of humanity. Miners sang at their work, at public gatherings, at festivals, at concerts, in church, and at communal parties. They sang ballads, minstrel songs, ditties, folk songs, and hymns, as solo items and in choirs.

Even plays and operas have been written in support of the joys and spiritual blessings of a miner's life. Every year during August a festival of miners' songs is held in the old mining town of La Unión, about ten miles east of Cartagena, Spain.

Musical bands have for centuries portrayed the miner's interest in music. At the many annual festival parades of miners in their ceremonial dress in various mining districts, the bands with their banners always lead the way. It was a poor sort of mining company that did not have its own band, especially in central Europe.

In Britain, to this day, the coal-mining communities are very proud of their mine bands. In fact, for the American Bicentennial celebration in 1976, two coal mine bands paid visits to the United States to perform on a concert circuit. These were the William Cory Band from Wales, and the Grimethorpe Colliery Band from Yorkshire. Both bands were of championship class. The Cory Band was Champion of Wales in 1966, 1967, 1969, and 1973, as well as being British National Brass Band Champions in 1974.

DRINKING VESSELS

Miners traditionally are hard-working joyful men; for relaxation, they drink ale as a unifying social custom. This does not infer insobriety, but a general predisposition toward conviviality and good fellowship. Among their treasured trophies and personal possessions, it is not unusual to find a drinking vessel figuring as a centerpiece.

From the 16th century on, many metalsmiths and artisans have designed drinking vessels as the central theme of their artistic endeavors; some beautiful examples are still held in many musuems of Europe. Each particular drinking cup has its own historical connection. Some fine examples are displayed pictorially in color in the halls of the department of mining engineering at the Virginia Polytechnic Institute and State University (Blacksburg, Virginia), through the courtesy of the mining firm Westfalia Luenen.

One such beautifully executed example is the Silver Tankard of the Harz Union (see plate 23.1).

> This silver tankard, made of Harz Silver, was designed by the chief inspector of mines, Von Imhoff, and produced by the goldsmith, H.A. Schumacher, of Wolfenbuettel. The engravings were executed by Schmidt, an engraver from Brunswick.
>
> The 40 cm high tankard is a cylindrical vessel with ear (handle) and lid. It has a weight of 5 kg and a capacity of four liters. An eyrie of red silver mineral specimens is from Andreasberg, with a griffin standing in it on a rectangular socle and holding a gad in his raised claw and a scroll in the other with the words: _Haec omnia munera Jovis._ Two golden medallions are set laterally in the lid border, one showing the head of King George II of Hanover, the other a portrait of Duke Augustus William of Brunswick. A third medallion of silver shows the Guelph steed surrounded by laurel sprigs. These three medallions signify that the Rammelsberg and the Communion Upper Harz are owned in common by the two Guelphic lines of Hanover and Brunswick. This connection is also stressed by a fourth silver medallion bearing the inscription: _Concordia in communione._ The plate of the lid support shows the sun whose golden rays were said to exert a great influence upon the formation of metals, and especially of gold.
>
> The outer wall of the tankard is adorned with three large oval and five small circular engravings. One oval shows men working on a ditch, water engineering being most significant for the Harz mining industry. The second oval shows an Upper Harz ore vein, with ladder, pump and hoisting shaft, as well as miners working ore. Above ground, one sees rods running from the water wheel housing to the shaft. On the right there is a stamp

Plate 23.1. The Harz Union Tankard.

mill and some dressers working at the slime table. In the background, there is the town of Zellerfeld, the official residence of the Harz Union Mine inspectors; and the Zellerfeld Mining Office is also to be seen. In the third oval, fire-setting is illustrated. It was still practiced at that time in shrinkage stoping. Above ground, one can recognize the "Maltermeister Tower" and several whims; on the right side, the Julius Smelting Works are outlined. The five small engravings symbolically illustrate five metals in the Upper Harz and in the Rammelsberg: gold is symbolized by Apollo, silver by Diana, copper by Venus, lead by Saturn, and iron by Mars.

On the foot of the tankard, the coat of arms of the Rammelsberg Mining Office is shown beside the miners' greeting "Glueck auf!" as well as the coat of arms bestowed on the Zellerfeld Mining Office in 1570. In between there are some mining motifs.

The ear (or handle) of the tankard is formed by a female figure with seven breasts which signify the seven dynasties forming the Harz Union and the profit obtained by this harmonious arrangement. On her head the woman carries a crown representing a crenulated wall. Deer and cattle are engraved on rings arranged around her hips and upper thighs, symbolizing the fundamental food supply in the Upper Harz.

The tankard represents a masterpiece of the goldsmith's art of that time. According to an old custom, the tankard is used for an all-round drink after the annual balance sheet has been reported, as well as on other festive occasions. The chief inspector of mines enters the hall preceded by the Miners' Band, the tankard filled with Rhine wine in his hands. With a friendly harangue, he presents it to the guest following him in rank order. This guest has to recite a mining saw and then take a gulp of wine. Thereafter, the tankard is passed around among the other guests. On the inside of the lid a toast is engraved as a verse expressing good wishes for the welfare of the Harz Union.

The original of the Silver Tankard of the Harz Union is owned by the Preussag AG Company in Goslar.

THE MINTING OF COINS

In Anatolia, after 1800 B.C. silver discs called shekels were first used as a unit of weight. In 680 B.C. coins were stamped upon discs of electrum metal (a natural alloy of gold and silver) by the Lydians. This was the first recorded use of coinage as a form of currency to replace the barter system. In any country, the minting of coinage for commercial currency was largely dependent upon the strength of the mining industry.

Joachimsthal, Freiberg, and other mining cities of the Erzgebirge possessed their own mints. Coins were minted and circulated with a value corresponding to the actual weight of contained silver, although each particular mint had its own currency standards.

In the Harz Mountains, coins were minted in various denominations of "loesers". The unit of currency employed at Joachimsthal was called the Joachimsthaler, or thaler for short. Coins were issued in several denominations of thalers. It is from the thaler that our term "dollar" is derived. The earliest silver thaler (dollar) was minted in the year 1519.

Inscriptions on coins usually bore the effigy of the local ruling prince, or his coat-of-arms, on one side, with a mining scene on the other. A description of the two-dollar loeser follows (see plate 23.2): On its front this thaler (two-dollar

Plate 23.2. A Julius loeser.

coin) shows the lower Saxony steed; above it, the arm with a laurel wreath being stretched forward out of the clouds in order to symbolize the blessing over the mining activities represented by individual scenes. These scenes are masterfully depicted on the small space available: the dowser, the prospector, the wheelbarrow man above ground; two horse whims with power transmission by pull-rods from the water wheel housing at the right side below; the kibble hoist underground; the windlass man; and the miner working at the face with hammer and gad.

The introduction of these two- or three-dollar coins dates back to Duke Julius the Lucky who staunchly supported mining in the Harz. He ordered that citizens of his duchy should buy a number of these Julius loesers, according to their financial position, with the understanding that they were not to be spent. In this way, he collected a hidden financial reserve for times of need, and work at the mines was also thereby encouraged. The Julius Tower, built for war funding in Spandau, was named after him.

The original two-dollar Julius loeser belongs to the coin collection in the possession of the Oberharzer Heimatmuseum (Upper Harz Museum) in Clausthal-Zellerfeld. By 1873, the terms *thaler* and *loeser* were superseded in Germany by the *mark*. In 1787, the *dollar* became the unit of currency in the United States.

24 Mining Customs

THE UNIVERSAL SYMBOL OF MINING

Just as other learned professions have their emblems, the universal symbol of the mining profession is the hammer and gad (or, in German, the Schlaegel-und-Eisen). The gad is a type of miner's chisel fitted with a handle. It is struck on the rear end with a hammer to loosen ore in the face (see figure 24.1). The crossed tools (hammer and gad) appear to have originated in 1420 in the old free mining city of Kremnitz, in the Ore Mountains of Slovakia. This city was founded by the Saxons in the 8th century. It became a free mining city in 1328.

Fig. 24.1. Hammer and gad.

Another possible source of its adoption is in the town of Zeiring, in Upper Styria, Austria. An early style of the emblem appears here, in an old miners' church, built in the year 1111. The emblem was probably set in the church in 1361 when the Zeiring silver mines were abandoned due to a terrible inrush of water.

As the firmly established symbol of the mining profession, this emblem is used all over the world in various ways: on mine buildings, at the entrance halls of mining schools, in churches, and on publications, documents, uniforms, badges, seals, altars, and in museums. The worldwide fraternity of mining engineers is linked and bound by this emblem. It usually forms the centerpiece of each national institute of mining.

THE MINER'S UNIFORM

In the cultural study of mining folklore, the miner's garb is of particular interest because its design and manner of wearing are deeply associated with many of the customs that have developed. The ordinary garment of the working miner had special parts peculiar to his work functions. It was generally a simple black smock with a pointed hood that could be pulled over the head to keep the hair free of dust, drips of water, and falling pieces of rock. With the smock, short breeches and shoes with gaiters were worn, as well as knee pads. This design gave the miner mobility for the many crawling, stooping, and crouching positions necessary in his daily work.

It is interesting to note that the black smock, leather belt, pointed hood, and long beard have been preserved in the corresponding attire of the Seven Dwarfs. Actually, the clothing worn by dwarfs or gnomes was none other than a representation of medieval miners' garb although in even earlier times the garb was white, especially in the salt mines of Hallstatt, Austria.

Several great traditions have developed from the standard miners' garb of Germany and eastern Europe. One of these is the arschleder, or arse leather. This leather apron was fastened with a leather belt around the waist of the miner, but with the apron at the rear to protect the buttocks while sliding over rocks and while sitting on damp ground. By custom, the arschleder became recognized over many centuries as a special distinctive symbol in the dress of the miner. Only fully fledged, experienced miners were entitled to wear the arschleder. They were in a class known as "miners of the leather."

One fascinating custom that developed from acceptance of the arschleder was the Ledersprung which means the "jump over the leather apron." This is an old initiation ritual, practiced in Austria and many of the old provinces of present-day Czechoslovakia and Yugoslavia. The young mining initiate or mining engineering graduate stands on a chair and the apron is held by its straps by two officials, one on either side, at some level above the initiate's feet. He must then jump, with both feet together, over the apron and make an elegant landing on the floor, in the manner of an accomplished gymnast. The action of the Ledersprung symbolizes the initiate's entry into the noble world of the mining profession.

Since the 17th century, many immigrant miners have arrived in Germany. The new arrivals were not so eager to embrace the old traditions, and some of these have since gradually waned. Nevertheless, the festal version of the garment is still honored by custom, not only by mining officials and professors, but also by the general body of workers. Each country has its own particular design of festal uniform, but generally it is a pleated frock-cut suit, black or olive in color, and embroidered in gold with gilt buttons and special symbols relating to the rank and style of the official, whether in mining or metallurgy. These uniforms are very colorful. They lend an aura of tradition to the particular functions at which they are still regularly worn, whether it be a divine service, a parade, a dedicatory observance, a funeral procession, a wedding, a professional conference, or a graduation ceremony. The members of miners' musical bands wear resplendent uniforms in the same general model.

In all the different designs of miners' festal uniforms according to whatever rank, one central feature remains invariant. They all incorporate the universal symbol of mining, the crossed hammer and gad, woven in gold on the upper shoulders and on the front of the cap, with the same insignia on the buttons. Associated with the festal uniform of mining officials was the berghaeckel, an ornamental metal blade fixed in a handle, as a sort of staff or mace.

A unique collection (for the United States) of framed water-colored prints of all ranks of mining officials in their respective festal uniforms is displayed in the hall of the College of Mines at the University of Idaho. In a book published in Nuremburg in 1721, Christoph Weigel describes the duties of various mine officials and shows 49 full-page, hand-colored pictures of these officers in their festal uniforms (parade dress).

THE MINERS' GREETING

The German miners' greeting is: "Glueck auf!" There is no precise translation available in English because it has a metaphorical connotation. Perhaps it can be interpreted as "Good luck" or "God speed to you!"

"Glueck auf" is used traditionally as a greeting between miners (and by most inhabitants of mining districts), as a toast in mining communities, as a motto, as an introduction to mining announcements, as a closing salutation in letters of correspondence, and generally in many ways to establish one's proud membership of the mining profession. There is a monthly technical mining journal published in Essen under the name Glueck auf.

This traditional greeting is supposed to have originated in the Saxon-Bohemian Erzgebirge in the 16th century. Many poems and songs have been composed about "Glueck auf." It can be regarded as a rallying call among miners. It seems to epitomize the love and spirit of a miner's life.

25 Mining Folklore

MINIATURE DENIZENS OF THE MINE

In medieval times, when supersitition and supernatural beliefs were rife, many mines were commonly believed to provide a habitat for little people, or wee folk. These were referred to by various names such as dwarfs, gnomes, sprites, pixies, demons, ghosts, trolls, goblins, elves, knockers (in Cornwall), and stope owls (in North America). These small non-humans played a significant part of mining folklore. The dark mysterious confines of a mine provided a suitable atmosphere in which imagination and superstition could be developed.

These little folk can be classified into two main types. First, the gentle kind, as observed by Greek and German miners. Only about two feet high but with adult and even aged facial appearance, they were usually dressed in miner's garb. They have been described as outwardly busy with simple miners' tasks, but they actually accomplished little. They gave no trouble to the miners but teased them by throwing pebbles at them, while mimicking them and gleefully laughing all the while. These little gnomes seemed to appear in stopes of high-grade ore, or where the prospects of finding such ore was good. They stimulated and encouraged the miners. They loved music, feasting, and dancing; they were also fond of playing mischievous tricks. However, if offended, they were ready for revenge.

On the other hand, the goblins or demons were an evil type: cruel, wicked, and black in the heart. These were the ones that appear to work behind the scenes creating dangerous situations for the miners: doing the devil's work! Agricola writes that miners were so badly attacked in the Rosenkrantz Mine at Annaberg that 12 were killed, and this rich silver mine was accordingly abandoned.

Later, in the mines of Cornwall, the "tommy-knockers" existed. These little folk teased the miners and mimicked the sound of their picks. Occasionally, pebbles would be tossed or food would be begged, under threat of bad luck for the morrow. They could be vindictive, and the miners always thought it better to leave a crust. Knockers had large, ugly heads with a withered appearance, some with long, gray beards. Cornish miners believed that knockers were the souls of departed spirits, drifting between heaven and hell, and wandering about the mines.

In Sweden, evidences of supernatural, ghostlike appearances in the stopes are recorded in a story entitled The Mines of Falun, one of the Tales of Hoffman. As far as North America is concerned, dwarfs and gnomes were not recorded until the arrival of Cornish miners following the California gold rush of 1848. Knockers tapped on the rocks and generally made their presence felt. Similarly, ghosts appeared and even spoke. Sometimes a ghostlike hand appeared, clutching a candle; but otherwise, they were completely invisible. The shrill cry of the "stope owl" has often been heard in mines on the North American continent. It was believed to presage a warning of a stope collapse or some other disaster.

But what of the presence of these supernatural beings today? Have they become "extinct"; or do they go unnoticed because their speech and sounds are drowned by the intensive noise from present-day machinery? Or perhaps their shadows are now nonexistent in these days of much superior underground illumination? Then today, miners are perhaps less likely to be so superstitious as to accept the presence of gnomes, dwarfs, and demons.

It is interesting to note that at Salzburg, center of an old mining district in Austria, statues of the Seven Dwarfs appear in an attractive setting in the Mirabell Gardens.

WOMEN IN THE MINE

In many mining communities, even today, there is a well-known belief that women should not be allowed to enter an underground mine. There seems to be no historical background to this rule. Only three main reasons can logically account for such a prohibition. First, on the grounds of decency, because many miners discard their clothing when working in hot places, and they are prone to use coarse language; second, it was considered too risky to expose a woman to any untoward, unpredictable danger; and third, there seems to have developed some feeling that a woman's presence would bring bad luck to the miners.

In the face of all these possible prohibitions, it is known that women were actually employed at strenuous tasks in European coal mines up to the mid-19th century. From this time onward based on the findings of various boards of inquiry, the employment of women underground has been banned by law in Europe. It is interesting to note that, following representations by the Women's Liberation Movement in the United States in recent years, women have been judged to have the right to work underground; and some are presently so engaged in coal mines of the Appalachian region.

V
Epilogue

26 The Importance of Mineral Production to Any Nation

There are many lessons to be learned through history in all phases of human activity. In respect to mineral production, the conclusions that follow help to crystallize national objectives that may be pursued with advantage by both developed and emerging nations.

People in many countries do not generally realize the national importance of a healthy domestic mining industry. Yet the people of every nation should be more aware of one of the main cornerstones of their national well-being. Each nation should assign a high priority to the development and maintenance of an active and progressive domestic mining industry.

All mineral deposits occurring within the borders of any country represent latent or potential national wealth to that country. They can be transformed into actual national wealth (and contribute to the gross national product) only by being mined.

Even though a comparatively few mine operating companies under the free-enterprise system are able to achieve large annual profits for their stockholders, this does not mean that the state receives no similar benefit. In fact, the state benefits in many ways: the gross national product is increased; employment levels are increased (on a multiplier basis); the level of self-sufficiency is enhanced; the balance of trade is improved due to fewer imports and greater exports of the commodity mined; a spirited search for more minerals is generated; technical manpower levels are built up by in-service training; overseas investment capital is attracted; and national wealth is created.

So not only do stockholders derive profit from their investment in a successful mining operation, every citizen of the state also gains indirectly. Furthermore, not by any means are all of the annual profits distributed to the stockholders; much of it is placed in the company's reserve fund to finance the search for or acquisition of new mineral deposits to replace eventually the one presently being depleted. Profitable operation is, therefore, of vital importance. Every nation needs to encourage mining companies to become profitable, in the interest of its own citizens.

It is, therefore, important for each country to encourage the development and extraction of its known mineral deposits, and to extract every available ton of mineral from each deposit as efficiently as possible. It is also to the advantage of each nation to encourage the active search for additional mineral deposits.

In advanced countries, the products of the mineral industry pervade almost every aspect of the life of every single individual. Until minerals were produced, way back in the Stone Age, the cavemen of the day led a very primitive existence. The subsequent development of mining and the use of mineral products has advanced civilization to its present status, with increasingly greater demands for these minerals over the years. Mineral production has provided the wherewithal for creating a better lifestyle for the individual. On the other hand, if domestic mineral production in terms of present needs begins to languish in the highly industrialized countries, then these people should be prepared to face a progressively worsening existence in the future.

As a typical highly developed country, let us look at America. The United States is regarded as the most highly industrialized nation in the world. Americans have been able to enjoy the highest of living standards because their country has been richly endowed with mineral wealth, and in the past these mineral deposits were exploited for their economic advantage. They were able to build up their great economic strength in this way because mineral wealth provides the means for moving developing countries toward a progressively higher standard of living, and continues to enrich countries already highly developed.

Accordingly, mining is an essential requirement for economic growth. Agriculture and mining (which includes extractive metallurgy) are the only industries that produce the basic raw materials with which all the other industries fabricate and service man's industrialized way of life. At the same time, the mining industry produces new national wealth. Mineral wealth is the backbone of any industrial society.

With few exceptions, no nation can enjoy a high level of prosperity without a reliable source of minerals to provide the necessary feedstock for its manufacturing industry. Through mining, many emergent countries can enhance their rate of growth by the export of raw mineral resources. The next step is to process these raw minerals prior to export. This has the effect of increasing local employment and adding value to the materials exported. At a later stage, these countries could develop their manufacturing capacities, based upon their existing mineral production activities. This leads to a progressive state of industrial development, the pace of which will necessarily depend, however, upon their command of technologically trained manpower.

Again, for every worker employed directly in the mining industry of any industrialized nation, at least five others are sustained in indirect employment to support its maintenance, equipment, transportation, and other needs. Mining is thereby able to provide jobs directly or indirectly for a substantial percentage of the non-farm workers of any nation. Most certainly, mining has a significant multiplier effect on the economy.

THE IMPORTANCE OF MINING TO THE INDIVIDUAL CITIZEN

Throughout history, man has found it necessary to dig in the earth's crust for materials to cook his food, to make his tools and weapons, and to heat his home. These same basic needs still exist today, and additional needs have developed as the standard of living has increased.

But in today's complex modern society, most people tend to take for granted the availability of electricity, hot water, refrigerators, automobiles, airplanes, calculators, books, newspapers, radio, television, kitchen gadgets, and all manner of plastic goods. These are only part of the myriad array of goods and services used in our daily lives without a second thought as to their origin.

These products are developed from the minerals produced from the earth's crust: from natural resources such as iron ore to provide the steel for automobiles, refrigerators, filing cabinets, kitchen utensils, razor blades, and the steel frames or reinforcing steel rods for buildings, shops, homes, schools, and bridges; also bauxite ore for aluminum window frames, light-weight car engines, kitchen utensils, beverage cans, and airplane bodies. Transport systems also involve large quantities of copper, lead, zinc, nickel, and other metals in their construction.

The generation of electricity and its distribution around the streets of any city (and between cities) call for large quantities of copper, aluminum, other metals, and insulating materials. Water pipes in large city networks are made from cast iron, steel, copper, and reinforced concrete. Similarly, the fuels used to heat homes and offices, to make steel, and to generate electric power are all recovered from the earth by some type of mineral production operation.

Wherever one works - in an office, in a classroom, in a kitchen, in a store, in a library, in a factory, or on a farm - one can observe a large number of items used each day that are produced by mining activity. It does not take much thought to realize that all the materials and conveniences so essential to daily life in an advanced country come from the earth directly or indirectly in one form or another, whether items of food, clothing, housing, transportation, communications, health and safety, or general comfort.

THE NEED FOR SELF-SUFFICIENCY IN MINERAL PRODUCTION

Minerals are the backbone and lifeblood of the economy of any nation. They are the basic materials from which substantially all other processed products are derived. A ready supply of mineral materials is, therefore, of paramount concern to the industrialized nations, and is a necessary condition for underdeveloped countries to "emerge" successfully.

No country is completely self-sufficient in that it can secure all of its raw material mineral needs from its own domestic deposits. Some countries more nearly approach this ideal than others. Obviously, there would be very great national advantages in being completely self-sufficient. It should be the policy of every country to aim for this objective. It not only reduces problems of political dependence, it also promotes national security and reduces international balance of payments problems.

Future prospects for an adequate supply of minerals for the world's industrial needs are not clear. As countries grow industrially, they compete for the world's output of mineral materials. And all nations naturally seek to upgrade their raw materials before offering them on the export market, necessarily at much higher prices. The competition for the available supply of critical minerals in the world is increasing year by year. As a result, metal and mineral prices are escalating, thereby contributing to economic inflation worldwide. It is apparent that every country should take steps to improve its state of self-sufficiency in mineral production. This can be done only if its general public begins to understand the vital importance of minerals to their existence and to their living standards. Only then will they be able to bring pressure to bear for the necessary legislation to improve domestic self-sufficiency in mineral production.

27 Financial Aspects of Mineral Development

Now that we know a few of the lessons of history as applied to the economic well-being of free-enterprise nations, let us analyse some of the forces involved in establishing domestic mineral production activities. Mining is a capital-intensive industry, much more so than most manufacturing industries. But mining is also a high-risk industry. Highly complex mining techniques are necessary to extract minerals from extremely variable ore deposits. At the same time, the prices received for the metals produced fluctuate widely over short periods of time in the world markets. Both these factors yield vagaries and uncertainties, giving rise to unpredictable financial results. These problems are accentuated as mining proceeds year by year to greater depths below the surface. The higher rock temperatures and pressures, and the greater drainage, ventilation, and extended hoisting demands associated with deeper working add materially to these problems.

In accounting terms, mines are regarded as "wasting assets" in that the mineral deposit on which the mine is based progressively becomes exhausted. At this point the mine necessarily needs to cease operations, unless in the meantime an extension of the deposit is found, or some other deposit is discovered, explored, and developed elsewhere. This exploration procedure can involve the expenditure of large amounts of capital dollars, always at a high degree of risk. The lead time necessary to explore, develop, and equip a mine for ore extraction varies from five to seven years or more. This program involves a considerable capital outlay before a single dollar is earned as revenue. In some cases, with mines in remote localities, a considerable proportion of this capital outlay is consumed in providing the necessary infrastructure, such as towns with employee housing, health and school facilities, amenities, highway and rail access, ports and airports, power and water supplies.

When the ore body becomes exhausted and the mine closes down, practically nothing is salvageable. Stockholders lose their investment. The annual dividends they have already received must be regarded not only as interest earned, but also as the partial return of their capital. For this reason, a mining company's operations cannot be regarded as successful unless a much higher return is provided than could be expected from an ordinary industrial investment, in order to provide a margin for the higher risks involved as well as for partial return of capital.

For all these reasons, it will readily be seen that there is a limited capital market for financing the high-risk mining industry. For every hundred mines financed to the point of production, only a very few become financially successful. This means that the ongoing provision of finance for opening new mines must be left mainly to the few successful mining groups that have been able to build up cash reserves by limiting the distribution of their annual profits.

Many people believe that the only function of a taxing authority is the gathering of taxes to feed government expenditures. But in many advanced countries, the tax mechanism is advantageously used as a fiscal tool to channel private investment into capital-poor areas. In this way, incentives are provided in the shape of specific tax deduction provisions. For instance, to encourage agricultural production, special tax provisions are generally allowed to farmers.

For mining companies, there is a real need to encourage high-risk investment by such special deductions; otherwise, domestic ore production would wane through lack of capital. It is unfortunate that the general public mistakenly regards these special tax provisions as "loopholes" that should be blocked. What is it that motivates entrepreneurs to engage in these difficult, risky enterprises? Obviously, as the late Professor H. W. Gartrell used to say: "Mining is not a legitimate field of investment for widows and orphans."

Prior to the introduction of income taxes in certain countries, some investors (not by any means all of them) made fantastic fortunes from mining ventures. This occurred during an era of extremely high-grade bonanzas, when operating costs were low, and when there was no tax liability. These profits supported a considerable interest in gambling (high-risk investment) in mining in those times. With the exhaustion of the richer ore deposits and the introduction of income tax regulations, the public interest in mining investment is now so low that capital-raising in the open market is difficult, borrowing or issuing of debentures is costly, and the only reliable source of venture capital is from a mining company's cash reserves, accumulated year by year from its undistributed profits.

Clearly, the number of such successful mining companies or groups is diminishing, and with currently escalating costs of establishing mines, their reserve funds are approaching extinction. In most developed countries, the trend is toward monopolistic control on the one hand, and stagnation of mineral production on the other.

In these days of difficult and costly capital funding, the levying of any form of taxes and royalties by the government of any country is counterproductive. On the contrary, the boot should be on the other foot. The government has much more to gain by not doing so. These principles have been clearly expressed in "The Gregory Formula" (Chemical Engineering and Mining Review, July 1958). If any country wishes to convert its latent wealth into actual national wealth (a most desirable aim), it should stimulate and encourage the accumulation of corporate profits in respect to mineral-producing operations, especially if economic nationalism is to be advanced. In the final analysis, this stimulation procedure could even amount to the foregoing of all taxes and the subsidizing of mining operations (in the general interest of the citizens), perhaps by progressive annual rebates of infrastructure costs from general revenue, after a specific number of years of successful, viable operation.

There is, therefore, a clear challenge for governments to encourage by appropriate fiscal measures the investment of greater amounts of risk capital to maintain or increase the rate of extraction of domestic mineral resources. Tremendous amounts of new capital will be required in the future to narrow the gap between domestic mineral demand and production in many countries. This is partly because presently known reserves of ore are of low or marginal grade, requiring special technology to render them economically feasible to mine.

28 The Impact of Mineral Production on National Economies

The reader will at this point have traversed the long trunk road and some of the many byroads of history in its specific relationship to the corresponding impact of mineral production on the cultural development of mankind over the ages. He will also have observed that many wars between tribes and nation-states have been fought either on religious grounds, or by various acts of aggression or invasion primarily to acquire mineral resources. It is patently evident that the development of national economies has always been indissolubly linked with the discovery of indigenous mineral resources and the development of mineral production capacity.

All through the ages, possession of mineral deposits and the ability to work and process minerals and metals have motivated tribes and nations either to develop their communities by raising their general cultural and living standards, or to defend themselves against invasion by others for the same general purpose. Since the Industrial Revolution, this objective can be translated into the achievement of a high level of industrialization, by whatever scale this can be measured, without any sacrifice of the quality of life.

Tangible examples stem from the Egyptian civilization (3300-300 B.C.), and from the Greek, Roman, German, Swedish, British, French, Austrian, and Spanish experience; and more recently from the American industrialization (following the California gold discovery), and the Australian, Canadian, South African, Mexican, Peruvian, Chilean, Brazilian, Venezuelan, Bolivian, Malaysian, Finnish, Romanian, Russian, and Chinese development; and very recently from the growth of OPEC countries. This is not an exhaustive list.

Of course, over the years there have been periods of slow growth or even of retardation due to a variety of causes, such as exhaustion of deposits (or insufficient capital resources to develop new deposits), competition from higher grade deposits abroad, political instability, excessive taxation, and short-sighted government interference and overregulation. It is also interesting to note that as the resulting industrialization proceeds to a level of comparative luxury, associated with a country's capacity to purchase its mineral requirements abroad, a general loss of community will to sustain or expand domestic mining activity sets in. This is the main cause of the current economic plight of Great Britain. Yet J. B. Richardson notes that the British Isles for their size are one of the most richly endowed with mineral deposits of any country in the world. During most of the 19th century, Britain was producing 75, 50, and 60 percent respectively of the world's copper, lead, and tin.

Among the many reasons for the decline in mining activity in Britain were excessive taxation, excessive distribution of profits as dividends to shareholders (instead of maintaining a viable corporate cash reserve), and lack of a studied government approach to a special taxation incentive. Meanwhile, the community is probably still reminded that the destruction of forests for charcoal production up to 1850 caused widespread damage to the environment. Perhaps the British community is prepared to forego or to reduce its standard of living rather than to restart an adequate level of mining activity. Several approaches (since 1950) have been attempted by companies to reopen old base metal mines, but apparently the community has resisted such moves. It wants to preserve this "green and pleasant land."

Meanwhile, apart from nonmetallic minerals and coal, nearly all metals are being imported, chiefly from former colonies. This represents a big economic loss. Can Britain continue to afford it? Or will another mineral, petroleum from the North Sea, be the savior of Britain's economy?

But what about other substantially industrialized economies? Will they have to face a similar down-grading of the economy through a community preoccupation with services, at the expense of support for and with a corresponding lack of interest in mineral production? Is America heading down this road? Will history repeat itself?

There are two major exceptions to the rule about the development of a high degree of industrialization resulting from an advanced level of mining activity (mineral production). One of these is Japan, which produces only a small proportion of its raw mineral requirements. Another is Switzerland, which produces little other than construction stone. How then, according to our hypothesis, can these two countries manage to rate among the world's leading industrialized countries?

The answer probably lies in the level at which their manufacturing industries are operated. Landlocked Switzerland produces high-precision instruments and high-precision machinery ranging from delicate wristwatches to the largest steam turbines, pumps, and marine diesel engines; for all of which it enjoys an extremely high reputation in the world's markets. It can remain competitive because of these skills and the possession of ample low-cost hydroelectricity. With the latter, it can afford to process many mineral raw materials purchased abroad. Apart from these advantages, Switzerland has the world's safest banking industry and the world's most breathtaking scenery. These features attract both capital funds and a one-way tourist trade in a measure enjoyed by no other country.

On the other hand, Japan, an island nation with many adequate shipping ports, has built up its manufacturing capacity by sheer efficiency in regard to manufacturing technology and labor objectivity. In spite of this, Japan needs to import nearly all of its basic raw mineral materials, such as coal, iron ore, copper, nickel, and petroleum.

Apart from these two exceptional cases, the general message should now be evident. The development of a high standard of living in any country almost invariably depends upon the active search for, and exploitation of, indigenous raw materials. To encourage the finding and development of these resources, a statesmanlike fiscal incentive program must be established and maintained; and the export of these products, surplus to domestic needs, processed to a higher level of value, should be a very desirable national objective. As Rickard so aptly states:

> The deserts were traversed by hunters and traders, by soldiers and artisans, . . . but theirs was an empty conquest and a vain annexation until the miner spoke the word that unloosed the springs of human industry. He was not only the pioneer, but he left marks to show the way; be blazed the trail for civilization. He has done it with geographic exuberance and equatorial amplitude . . . as the herald of empire and the pioneer of industry. Trade follows the flag, but the flag follows the pick.

Appendix — Significant Dates in Mining History

968	Harz Mountains ore deposits discovered (Rammelsberg)
1000-1300	Mining revival at the end of the Dark Ages
1170	Discovery of rich silver ores at Freiberg, Saxony (Erzgebirge)
1300-1450	Slump in European mining caused by political instability (Hundred Years' War and Hussite Wars)
1451	Funcken developed a method of separating silver and copper
1450-1550	Expansion in European mining
1521	Conquistadores entered Mexico (Cortes)
1550-1750	Slump in European mining following entry of rich ores from the Americas
1550	Coal industry began in Britain
1556	Agricola published <u>De Re Metallica</u> in Latin
1627	Explosives (black powder) first used in mines (Slovakia)
1689	Explosives first used in Cornish mines
1702	Savery's atmospheric engine (forerunner of Cornish pump engine) Subsequent development of Cornish steam-driven pumps by various inventors: 1712 - Newcomen 1775 - Smeaton 1777 - Watt and Boulton 1811 - Trevithick
1744	Gold mining commenced in Russia
1770	Longwall coal mining first introduced (Britain) Industrial Revolution (Britain)
1815	Davy safety lamp introduced
1831	William Bickford invented safety fuse (Cornwall)

1842	Employment of women and children forbidden underground
1843	Copper mining commenced in Australia (Burra)
1844	Copper mining commenced in Upper Michigan
1848	Discovery of gold in California (gold rush)
1850	First mechanical rockdrill (Mont Cenis Tunnel, French Alps)
	Compressed air first used underground to replace steam (Cochrane)
1851	Discovery of gold in Australia (Ballarat gold rush)
1852	High-grade iron ore first produced in Lake Superior district
1859	The first American oil well (Titusville, Pennsylvania)
1860	Discovery of gold and silver (Comstock Lode, Nevada)
1860s	Discovery of gold in Rocky Mountains (Central City, Colorado)
	Discovery of gold in Queensland, Australia
1863	Alfred Nobel invented dynamite
1864	Rudolph Leschot invented the diamond drill (France)
1867	Discovery of diamonds in South Africa
1868	First successful coal cutter introduced
1870	Discovery of copper in Chile
1874	Discovery of gold in Black Hills of South Dakota (Homestake)
1878	Discovery of copper at Butte, Montana, and in Arizona
1883	Discovery of rich silver-lead ore at Broken Hill, Australia
1886	Gold discovered near Johannesburg, South Africa
1890	Fine coal dust first shown to be an explosive hazard
1892	Discovery of gold at Coolgardie (Western Australia)
1894	Gold discovered at Cripple Creek, Colorado (gold rush)
1897	Gold discovered at Klondike, Yukon (gold rush)
1900	Wet pneumatic rockdrill introduced (Leyner)
	Iron mines opened at Kiruna (northern Sweden)
1903	Bingham Canyon Copper Mine opened, Utah
	Rich gold strike at Cobalt, Ontario
1904	Flotation process invented at Broken Hill, Australia
1920s	Discovery of copper ore in Zambia and Zaire (Copper Belt)
	Discovery of lead-zinc ore bodies at Mount Isa, Australia
1950	Continuous miners introduced for underground coal mining
	Tungsten carbide hard metal introduced for drill bits

1950s	Discovery of iron ore in Labrador and Quebec
	Discovery of iron ore in Western Australia
1955	Blasting agents introduced (to replace dynamite)

Glossary of Mining Terms

ADIT A blind horizontal opening into a mountain, with only one entrance.

ASSAY The testing of a sample of minerals or ore to determine the content of valuable minerals in the sample.

BACK The ceiling of any underground excavation in a metalliferous mine.

BALL MILL An item of milling equipment used to grind ore into small particles. It uses steel balls as the grinding medium.

BASE METAL A commercial metal such as copper, lead, or zinc. The term is used to distinguish these metals from the precious metals.

BEDDED DEPOSIT A mineral deposit of tabular form that lies horizontally and is commonly parallel to the stratification of the enclosing rocks. A coal seam is a typical bedded deposit. Others may contain industrial minerals, and some are metalliferous.

BEDROCK The solid rock of the earth's crust, generally covered by overburden, soil, or water.

BENEFICIATE To treat mined mineral matter so that the resulting product is richer or more concentrated with the useful mineral.

CAGE An elevator-type conveyance that moves men and materials up and down a mine shaft by means of a hoist rope.

CALORIFIC VALUE The quantity of heat that can be liberated from one pound of coal or oil. It is measured in British thermal units (BTU). Coal, being a mineral of organic origin with no precise composition, cannot conveniently be evaluated in terms of grade. It is, therefore, evaluated in terms of its calorific value (CV).

CHUTE An opening in mine workings through which broken ore is moved into mine cars for haulage to the shaft.

CLAIM An area of land staked by a prospector or mining company and then recorded (see STAKING).

COAL RESERVES Measured tonnages of coal that have been calculated to occur in a coal seam within a particular property.

COLLAR The entranceway from the surface, or a level, of a shaft or winze.

CONCENTRATE To treat ore so that the resulting "concentrates" will contain less waste and a higher amount of valuable mineral. In many mining operations, ore is concentrated in a concentrator or mill on the surface, then shipped to a smelter for the recovery of metal (see BENEFICIATE).

CORE A cylindrical stem of rock that is extracted from the earth by a diamond drill. The core is removed to the surface for examination and analysis (assay).

CROSSCUT A horizontal opening driven across the strike of a vein, or across the direction of the main workings. A connection from a shaft to the vein.

CUT-AND-FILL STOPE A stope in which the ore is removed in slices, after which waste material (backfill) is run in before the next slice is mined. The backfill supports the walls of the stope.

DEVELOPMENT The work of driving openings to and in a proved ore body, or a coal seam, to prepare it for systematic mineral production.

DIP The angle at which the vein is inclined from the horizontal.

DRIFT A horizontal opening in or near an ore body, parallel to the strike of the vein, or the long dimension of the ore body. Otherwise, a type of coal mine in hilly country where the seam is entered horizontally (without the use of a shaft).

EVALUATION The work involved in gaining a knowledge of the grade, shape, position, and size of a prospect.

EXPLOITATION The work of mining and processing the ore for sale in the market.

EXPLORATION The search for a mineral deposit (prospecting) and the subsequent investigation of any deposit found (evaluation).

FAULT A break in the earth's crust caused by forces that have moved the rock on one side with respect to the other.

FERROUS Minerals that contain iron. Minerals of metals that do not contain iron are termed nonferrous.

FILL (or Backfill) Waste material used to support the walls of a stope and provide a working platform for the miners.

FLOTATION A commonly used milling process in which certain minerals in a water-borne pulp attach themselves to air bubbles and float to the surface, while gangue minerals sink to the bottom or pass out the exit, thereby causing separation.

FOOTWALL The wall or rock under an inclined vein. It is called the floor in bedded deposits.

GANGUE The mineral material in an ore, forming part of a vein, reef, or lode that is not commercially useful. Gangue minerals are discarded as tailings as soon as they can be separated from the useful or valuable minerals, during the concentration process.

GRADE The proportion of metal contained in unit weight of ore, even though the metal is in the form of a mineral. The grade is usually expressed as a percentage by weight. For instance, a grade of 4 percent lead means that a short ton of lead ore contains 80 lb of lead, usually in the form of the mineral galena (PbS). The ores containing precious metals contain minor amounts of the metal and cannot conveniently have their grades expressed in percentages. They are, therefore, expressed as troy ounces per short/long ton; or as grams per tonne.

HANGING WALL The wall or rock on the upper side of an inclined vein. It is called the roof in bedded deposits.

HAULAGE The horizontal transport of broken ore along a level to an ore pocket near the shaft.

HOISTING The vertical transport of broken ore up the shaft from the ore pocket to the ore bins on the surface.

INDUSTRIAL MINERALS Usually nonmetallic minerals that are used in industry and manufacturing processes in their natural state, though generally with some beneficiation to imposed specifications. Examples include asbestos, salt, potash, trona, and phosphate rock.

JAW CRUSHER A machine in which the ore is broken by the action of moving steel jaws.

LEVEL Mines are customarily worked from vertical shafts through horizontal passages (drifts and crosscuts) called levels. These are commonly spaced at regular intervals in depth, and are either numbered from the surface in regular order or designated by their actual distance below the collar of the shaft.

LODE A wide, near-vertical mineral deposit in a sheared zone of rock.

MILL A plant, usually at the mine site, that concentrates ore or treats it so that minerals are separated and prepared for ultimate recovery in a purer form.

MINERAL An inorganic compound occurring naturally in the earth's crust, with a distinctive set of physical properties and a definite chemical composition. Minerals are, therefore, assemblages of chemical elements. They may be described as "common rock-forming" minerals, or as minerals "of economic value." For convenience, native metals and organic hydrocarbon materials (like coal and petroleum) are also classified as minerals.

MINING The process of obtaining useful minerals from the earth's crust for the benefit of mankind. Mining is mineral production. It includes both underground excavations and surface workings.

MUCK Ore or rock that has been broken by blasting

OPTION A right to have the first chance to buy or refuse to buy a mining property.

ORE A natural aggregate of metalliferous minerals that can be extracted from the earth's crust at a profit. Ore contains commercially useful minerals as well as gangue minerals.

ORE BODY A mineral deposit containing ore, such as a vein or lode.

ORE DEPOSIT A metalliferous mineral deposit sufficiently concentrated by nature to warrant extraction by mining.

ORE POCKET An excavation in the rock near the shaft to store broken ore delivered by haulage trains, with chute gates feeding skips for hoisting to the surface.

ORE RESERVES Measured tonnages of ore of a certain grade in a particular deposit. May be described as proven, probable, or possible, depending upon the accuracy with which the size and grade of the ore has been determined.

OUTCROP The surface exposure of a mineral deposit. In some cases, where the uppermost part of a deposit is concealed by soil or overburden, there may be no outcrop.

OVERBURDEN Soil, plant life, or water that covers solid rock or a mineral deposit.

PORTAL The entrance to a tunnel or adit.

PROSPECT A mineral deposit, the value of which has not been adequately tested or evaluated.

PROSPECTING The search for a mineral deposit in the earth's crust.

RAISE A vertical or inclined opening driven upward from a level to connect with the level above, or to explore the ground for a limited distance above a level.

ROCK An assemblage of common minerals usually in random proportions (with no definite chemical composition). Usually, any useful minerals present in a rock are too dispersed to be of commercial value. But some rocks themselves are commercially useful, such as building stone, monumental stone, aggregate for concrete, etc.

ROYALTY Amounts of money paid by an operating mining company to the actual owner of the mineral rights to the property. The royalty may be based upon an agreed amount per ton, or a percentage of the revenue or profits.

RUN-OF-MINE Ore of average grade in a mine.

SHAFT A major development opening. A vertical (or inclined) excavation in a mine extending downward from the surface, or from some interior point, as a principal opening through which the mine is exploited. A shaft may be provided with a hoisting engine and headframe at the top for handling ore, men, and supplies; or may be used only in connection with pumping or ventilating operations, or to provide an escapeway. A shaft generally is divided into separate compartments.

SHRINKAGE STOPE A method of stoping that utilizes part of the broken ore as a working platform and as temporary support for the walls.

SMELTING The recovery of metal ingots from ore that has been treated and concentrated at a mill; smelting is required to recover the metal content in a form from which it can be sold or further refined before sale.

STAKING The measuring and marking with stakes or posts of an area on the ground to establish mineral rights.

STOPE The part of an ore body from which ore is currently being mined, or broken, by stoping (drilling and blasting).

STRIKE The horizontal course or bearing of an ore body, such as a vein. The direction of a horizontal line in the plane of a vein. The direction of a vein as it would appear on a horizontal land surface.

SUMP An excavation for the purpose of collecting or storing seepage water before pumping to the surface. The bottom of a shaft is sometimes used for this purpose.

TUNNEL A horizontal opening clear through a mountain, with two entrances.

VEIN A mineral deposit having a more or less regular development in length, width, or depth to give it a tabular form, commonly inclined at a considerable angle to the horizontal. The term lode is similar to a vein, but of much greater width.

WALL ROCK The rock forming the walls of a vein or lode.

WASTE Material that is too low in grade to be of economic value, e.g., broken barren rock or mullock.

WINZE A vertical or inclined opening sunk from a point inside a mine for the purpose of connecting with a lower level, or of exploring the ground for a limited depth below a level.

Bibliography

Abdal Atta, Abdal-Azinor. "Long Range Planning in the Sphere of Irrigation and Drainage." UNIDO Conference on Long Range Planning and Regional Integration, INP Translation (Cairo, Egypt: Egyptian Ministry of Irrigation, January, 1976).

Agricola, G. De Re Metallica, translated by H.C. and L.C. Hoover. New York: Dover Publications, 1912.

Annabell, R. The Uranium Hunters. Adelaide: Rigby Ltd., 1971.

Ashton, T.S. Industrial Revolution 1760-1830, rev. ed. London: OUP, 1964.

Ashton, T.S., and Sykes, J. The Coal Industry of the 18th Century. Manchester: Manchester University Press, 1964.

Auhl, I., and Marfleet, D. Australia's Earliest Mining Era. Adelaide: Rigby, 1975.

Avery, D. Not on Queen Victoria's Birthday: The Story of Rio Tinto Mines. London: Collins, 1974.

Barton, D.B. Mine Pumps. Truro: Barton, 1965.

Bennett, R.H. Quest for Ore. Minneapolis: T.S. Denison, 1974.

Bernstein, M.D. The Mexican Mining Industry, 1890-1950. Albany: State University of New York Press, 1965.

Biringuccio, V. Pirotechnica. Cambridge, MA: M.I.T. Press, 1966.

Blainey, G.N. The Rise of Broken Hill. Melbourne: Macmillan, 1968.

__________. The Rush that Never Ended, 3rd ed. Melbourne: Melbourne University Press, 1978.

__________. Mines in the Spinifex. Sydney: Angus & Robertson, 1970.

__________. The Peaks of Lyell, 4th ed. Melbourne: Melbourne University Press, 1978.

__________. Gold and Paper. Melbourne: Georgian House, 1958.

Boshier, A., and Beaumont, P. "Mining in Southern Africa and the Emergence of Modern Man." Optima, vol. 22, no. 6 (1972).

Branigan, K. Atlas of Ancient Civilizations. London: Heinemann, 1976.

Bryan, Sir Andrew. The Evolution of Health and Safety in Mines. London: Ashire, 1975.

Bryan, J.F. The First 50 Years of Independent Life. London: The Newcomen Society, 1950.

Butterman, W.C. Gold. Wash. D.C.: U.S. Bureau of Mines, MCP 25, 1978.

Camp, J.M., and Francis, C.B. The Making, Shaping and Treating of Steel, 5th ed. Pittsburgh: U.S. Steel Corp., 1940.

Carroll, B.M. Australia's Mines and Miners. Melbourne: Macmillan, 1977.

Chadwick, J. The Mycenaean World. Cambridge: CUP, 1976.

Coghill, I. Australia's Mineral Wealth. Melbourne: Sorrett, 1971.

Davies, D.C. Metalliferous Minerals and Mining, 4th ed. London: Crosby Lockwood, 1888.

Davies, E.N., and Northedge, G.A. Mining and Minerals. London: Pergamon, 1967.

Derry, T.K., and Williams, T.I. A Short History of Technology. Oxford: OUP, 1960.

Durant, W. and A. The Story of Civilization, vols. 7-10. New York: Simon and Schuster, 1967.

El-Hanafi, Saad el-Din. "The Sahpe of the Egyptian Agricultural Sector to the Year 2000." UNIDO Conference on Long Range Planning and Regional Integration, INP Translation (Cairo, Egypt, A.R.E., Institute for National Planning, January, 1976).

Flawn, P.T. Mineral Resources. New York: Wiley, 1971.

Food Agricultural Organization, Near East Regional Office, "Research on Crop-Water Use, Salt Affected Soils and Drainage in the Arab Republic of Egypt." (RAO, 1975).

Forbes, R.J. Studies in Ancient Technology, vols. 8-9. New York: Heinman, 1971-72.

Greever, W. St.C. The Bonanza West. Norman, OK: University of Oklahoma Press, 1963.

Gregory, C.E., ed. The Art of Mining in Rammelsberg. (private pub, 1969). This features the "Festive Oration" delivered by Oberbergrat Herbert Dennert on the occasion of the 1,000th Anniversary of the opening of the Rammelsberg mine.

_________. The Legend of St. Barbara, rev. ed. Blacksburg, VA: Virginia P.I. & S.U., 1975.

_________. "The Gregory Formula." Chemical Engineering and Mining Review. vol. 50, no. 10, (July 1958), and vol. 53, no. 10, (July 1961).

_________. "The Problems of the Mining Company." Chemical Engineering and Mining Review. vol. 50, no. 9, (June 1958).

Haldane, J.S. "Health and Safety in British Coal Mines." Historical Review of Coal Mining. London: Fleetwood Press, 1940.

Herm, G. The Phoenicians: The Purple Empire of the Ancient World. London: Gollancz, 1975.

Hillaby, R. A Walk through Britain. London: H. Mifflin, 1978.

Hogan, E. Gold Rush Country. Menlo Park, CA: Lane Book Co., 1976.

Hopper, R.J. The Early Greeks. New York: Barnes & Noble, 1976.

Hore-Lacy, I. Mining and the Environment. Sydney: Australian Mining Industry Council, 1976.

Jones, W.R. Minerals in Industry. Harmondsworth: Penguin, 1950.

Keesing, Nancy, ed. History of the Australian Gold Rushes. Melbourne: Angus & Robertson, 1976.

Kerrison, Janet. Beaconsfield Gold, 3rd ed. Launceston, Tas.: Rotary Club of Beaconsfield, 1976.

Kirnbauer, F. Der Ledersprung. Vienna: Montan Verlag, 1960.

Larrick, L.A. The Pierce Chronicle. Moscow, ID: University Press of Idaho, 1976.

Macqueen, J.G. The Hittites and Their Contemporaries in Asia Minor. Boulder, CO: Westview, 1975.

Marsland, L.W. The Charters Towers Gold Mines. London: Waterlow Bros. and Layton, 1892.

Morley, J., and Foley, Doris. Gold Cities -- Grass Valley and Nevada City. Berkeley: Howell-North Books, 1960.

Morrell, W.P. The Gold Rushes. London: A. and C. Black, 1940.

Morrison, T.A. "Some Historical Notes on Mining in the Harz Mountains, Germany." Trans. Instn Min. Metall., Sect. A, April 1976.

Nef, J.U. Rise of the British Coal Industry. Chicago: Chicago University Press, 1966.

Park, C.F., Jr., and Freeman, Margaret C. Earthbound. San Francisco: Freeman Cooper, 1975.

Paul, R.W. California Gold: The Beginning of Mining in the Far West. Lincoln: University of Nebraska Press, 1965.

Paul, W. Mining Lore. Portland, OR: Morris Printing Co., 1970.

Pearl, R.M. Rocks and Minerals. New York: Barnes & Noble, 1956.

Pohs, H.A. Early Underground Mine Lamps. Tucson: Arizona Historical Society, 1974.

Poss, J.R. Stones of Destiny. Houghton, MI: Michigan Technological University, 1975.

Prain, Sir Ronald. Copper -- The Anatomy of an Industry. London: Mining Journal Books, 1975.

Prieto, C. Mining in the New World. New York: McGraw-Hill, 1973.

Reunert, T. Diamonds and Gold in South Africa. London: Books for Libraries Press, 1893.

Richardson, J.B. Metal Mining. London: Allen Lane, 1974.

Richardson, W.G. A Survey of Canadian Mining History. Montreal: Canadian Institute of Mining & Metallurgy, 1976.

Rickard, T.A. Man and Metals. 2 vols. New York: McGraw-Hill, 1932.

Rogan, J.M. Medicine in the Mining Industries. Philadelphia: F.A. Davis Co., 1972.

Sealey, R. A History of the Greek City States, 700-338 B.C. Berkeley: Univ. of California Press, 1977.

Sinclair, J. Environmental Conditions in Coal Mines. London: Pitman, 1978.

Sloane, H.N. and L.L. A Pictorial History of American Mining. New York: Crown Publishers, 1970.

Smith, G.H. The History of the Comstock Lode. Reno: University of Nevada, 1970.

Spence, C.C. Mining Engineers and the American West. New Haven: Yale University Press, 1970.

St. Clair, H.W. Mineral Industry in Early America. Wash. D.C.: U.S. Bureau of Mines, 1977.

Stewart, J.M. "Siberia -- No longer a Sleeping Land." Optima, vol. 26, no. 2, (1976).

Stora Kopparberg, A Chronicle from the Beginning to the Present Day. Stockholm: Stora Kopparbergs Bergslags AB, n.d.

Takeuchi, H; Uyeda, S; and Hanamori, K. Debate about the Earth. San Francisco: Freeman Cooper, 1970.

Temple, J. Mining -- An International History. New York: Praeger, 1972.

The Burra Record, vol. 61, no. 17. April 16, 1940.

The Importance of Minerals. Wash. DC: U.S. Bureau of Mines, 1974.

Thomas, L.J. An Introduction to Mining, 2nd ed. Sydney: Methuen, 1978.

U.S. Department of the Interior. "Success at Oil Creek." In Historical Vignettes, 1776-1976. Wash. DC: USDI, 1976.

Wilson, A.J. "Timna's Ancient Mining Secrets." Optima, vol. 26, no. 1 (1976).

Winkelmann, H. Der Bergbau in der Kunst. Essen: Glueckauf Verlag, 1960.

Young, O.E., Jr. Western Mining. Norman, OK: University of Oklahoma Press, 1960.

Zola, E. Germinal. Harmondsworth: Penguin, 1954.

Index

About the Author

Cedric E. Gregory was born in Adelaide, Australia and received his first degree in mining engineering at the University of Adelaide during the Depression. Subsequently, he deliberately sought practical employment as an underground hardrock miner to prepare himself for professional life. Soon after, he became a senior executive in the management of mining operations. Following four years of World War II duty with the Royal Australian Engineers in the Pacific, he entered industry as a secondary career and occupied senior positions in industrial management until he suffered a severe accident which required two years of recovery.

After his convalescence, he entered academic life as a university instructor. This proved most satisfying because his extensive background of practical mining experience brought to the classroom a rich supplement to textbook theory.

Gregory worked hard to enhance his academic status, devoting six years of personal research to a crucial aspect of mine ventilation for which he subsequently earned an international reputation.

Dr. Gregory retired as Professor Emeritus of Mining Engineering from the University of Idaho in 1974, and made preparations to live quietly on the Mediterranean coast of Spain. However, he has since been called out as a Visiting Professor at Virginia Polytechnic Institute and State University and at King Abdulaziz University in Saudi Arabia.

His major hobbies are world travel, writing, inspiring students, and attending international conferences, activities in which he is ably and devotedly assisted by his charming wife, as it has been throughout their 45 years of union. They especially enjoy periodic visits to Australia to visit with their five children and 12 grandchildren.